MW01380020

This book is a gift from the
Range Conservation Foundation
made possible by funding from the
Laura Moore Cunningham
Foundation

Nippers and Oldies

The Long Trail Home

Early days 'til now

Published by Range Conservation Foundation & RANGE magazine
Produced and edited by C.J. Hadley

Photo: Winter cattle drive at the Pitchfork Ranch in Meeteetse, Wyoming.
By Charles J. Belden (1887-1966)

Ready to Roll

"I'm not like other kids.
I don't need to go to school, because
I already know everything."

*Billy DeLong, age five, on the first day of homeschool
on the Black Rock Desert.*

PUBLISHER/EDITOR: *C.J. Hadley*
ASSOCIATE EDITOR: *Carolyn Dufurrena*
DESIGNER: *John Bardwell*
PROOFREADER: *Denyse Pellettieri White*

*"Nippers and Oldies:
The Long Trail Home"
Hadley, C.J.
PCN 978-0-578-98444-5
Library of Congress Control Number:
2021918088*

Published by Range Conservation Foundation & RANGE magazine, Carson City, Nevada. All rights reserved.

*$38 U.S.A.
Printed in South Korea
Copyright © 2021
Range Conservation Foundation
& RANGE magazine*

Matti, age two-and-a-half, with big brother, Billy, ride the drag. They are helping to move cattle near their family's Jackson Creek Ranch in Nevada's Black Rock Desert, 90 miles northwest of Winnemucca. (See page 90) Photo © Katie DeLong

A Century of Memories

By C.J. Hadley

This book is filled with hardworking, courageous independents who settled tough country to build their dreams. The stories cover more than a century and the stars range from a few days old to 103. (The dates these tales originated in "Confessions of Red Meat Survivors" in *RANGE* magazine are noted on many.)

Page 12: "One year, the Easleys paid their taxes with money received for bounty on jackrabbit ears."

You will find stories of ancients who have lived and worked through several wars, the Great Depression, fickle federal rules and regulations and anti-grazing activists who envy their self-sufficiency and want them gone.

Page 110: "We had a little starve-out place. I always had a saddle horse, but I didn't always have a saddle."

Often tired and hungry, always laboring, they built a beautiful world from pretty much nothing. They cherished their families, loved and fought Mother Nature, and even if a real payday took decades to arrive, their satisfaction came from the value of the work.

Page 80: "I was told I had to go to school to learn how to read, but after my first day at school I came home and picked up the paper and still couldn't read!"

Most within these pages come from ranching. They span multiple generations and tiny cowboy earthlings now mimic the old ones, take on chores with gusto, and have love in their hearts for family, country, livestock, horses, wild things and dogs.

Page 41: "It was eight years of root, hog, or go hungry. And then four kids sat at our table. Looked like a bird's nest with four little open mouths waiting to be stuffed."

These miniature humans stand on the shoulders of their ancestors, learn from the wise ones and follow their lead.

Page 130: "Dad and the other modern-day pioneers domesticated this wild place from its prior denizens, the rocks, rattlers, and sagebrush."

These valiant souls are doers and thinkers. They covet nothing that they haven't earned and sacrifice for family and the chance of owning a ranch filled with cattle or sheep that help feed and clothe the world.

Page 55: "Life was rugged. Jim and Philomine sometimes lived in a tent with a wood floor or a small cabin built atop a wagon."

"Nippers and Oldies" is a glimpse at the future and at the past, a reminder of what America was and could still be—an infant country based in liberty that learns and improves as it grows.

Page 19: "The scarcity of water was constant. Sam rigged up a sled, put a 50-gallon water barrel on it, and hauled water six miles from our windmill down a rocky canyon. He gave me a glass of water in the morning and said, 'You can either take a bath in it or drink it.'"

When asked why they work so hard for so little, many ranchers say that there is nothing they would rather do. Maybe they know that food producers are the most important people in the world and enjoy some satisfaction from a life well lived, well earned and well done. And they know they have something of value to pass on with a smile to the little ones. ∎

A Cowboy Romance

Green-eyed buckaroo meets flustered town girl with chicken blood on her saddle shoes. "I thought she was pretty cute before that." Bill and Nita Lowry, 89 and 86, have ridden a rocky trail for 73 years on their dream place in the Owyhee rimrocks. (See page 108) Photo © Carolyn Dufurrena

Looking Back, Looking Ahead

By Carolyn Dufurrena

Little Wrangler

My grandson, Ivan Acciari, wanted to go for a ride, so I asked him to catch my horse, Spotz. At the time he was two-and-a-half years old and thought all he needed was a black hat and a good rope. © Cathy Acciari

The trail begins sometime in the last half of the 1800s. They come from Spain, from Austria, from Germany, Ireland, and Norway. Basques come through Ellis Island or up the cattle trails from Mexico. They leave poverty and hardship and look west to find hardship—but maybe not poverty.

They follow their ancestors and bring their children, their children's children, a niece, a nephew. The young ones play along the way, learn the tools of the trade, stepping in when the old ones need a hand. Buckarooing long before the fences, settling in the Sandhills of Nebraska and the Missouri Breaks of Montana, they live a life outside, with horses and ladies in woolly chaps.

They grow up in flour-sack bloomers and hand-me-down overalls, work as one-room schoolteachers and fill-in cooks for the ranch crew. They take the kids to school in a team-drawn bobsled when the snow is deep, and sometimes the kids do the hard, outside work alone because that's all the help there is.

They pack salt on mules into the wilderness for wild game, pack milk cans full of trout into mountain streams in the Idaho wilderness, hitchhike to Iowa to pick corn, herd sheep and run cattle in the middle of nowhere. They fight prairie dogs, pack rats, beaverslides and stackers, herd turkeys, check windmills, put up hay and deliver mail.

Times get better after the war, with more opportunity. More sheep, then more cows. They run mustangs with airplanes, get his and hers tractors that they label "Jessie" and "Carl." They build bridle reins, horseshoes, even necklaces of rattles. They still fight disease: hoof and mouth, scabies. The sheep still get lost, and later are found.

They raise kids on places called Swamp Angel and Crooked Creek, the Grindstone, Last Chance and Temperance Creek,

Just a Few Great-Grands

Evelyn Penrose Cuneo came from a family of Cornish miners who migrated from Michigan to Jackson, California, in the 1920s. They reinvented themselves as ranchers and dairy farmers. Evelyn's boyfriend Vernon Cuneo took her horseback once, on a 30-mile cattle drive. It was her first and last time on a horse. Nonetheless, they married, raised five children on the family ranch, and celebrate holidays there with grands and greats. This shows Evelyn with nine of her 12 great-grandchildren on July 4, 2012, at Cuneo Camp near Blue Creek in Calaveras County. Back row: Brady Blake, Michael Blake, Della Lefty and Dottie Blake. Front row: Jada Payne, Jacob Payne, Carly Payne (on Evelyn's lap), Adain Flynn and Jack Flynn. Jada is 10 months old; Evelyn is 93. (See page 58) Photo © Larry Angier

Buenos Aires, Calcutta, Wolf Creek and Padlock, the IL, YP, Raft River and LO.

Great-grandpas deliver their granddaughters' babies, bury their own kids sometimes. They are tough—tough by nature as well as experience. They have "a sense of humor to avoid snakes, skunks and the occasional runaway tractor," as one says.

That and "a daily shot of vodka with raisins to fend off arthritis," or whatever will help get you and your little ones on down the trail.

You can only see the trail till the next turn, but these old ones have been here before. They know the way, and they urge the youngsters along. Some trails are

rough. Some are steep. Some are so faint they barely exist. Few are straight enough that you can see the destination, and maybe you wouldn't even want that. The bends in the trail bring adventure, the unknown, and the opportunity for caring for each other that bring us down the long trail home. ■

Contents

Safety Zone
At one time or another every ranch kid gets sent to the truck when a grandparent or parent decides a cowboy situation is temporarily unsafe for tykes. Such was the case at the Home Ranch in Jackson, Montana, while bringing the cattle in for sorting. Kaleb Wayne Donker, four, seems to be enjoying his truck time where evidently snacks are available. (See page 143) Photo © Cynthia Baldauf

"Everybody's outdoor sport was cattle rustling."
Beth Smith Aycock, age 95, in Green Valley, Arizona. (See page 19)

(Continued on next page)

With two wagons,
five young children,
and a small herd of cattle,
these strapping
servants of the Lord
headed east into Utah's
terra incognita.

At 91, never ready to quit, Sara Ann Holyoak has few cares and enjoys the company of family who revel in her colorful tales of Moab's bygone years. (See page 136)

Young Gus Melin enjoys a drink of milk in the saddle while watching his family brand cattle on their ranch near Pray, Montana.
© Todd Klassy

At seven, Jack learned to trap and hunt muskrats, skunks and coyotes. "I used the money I earned from hides to buy .22 shells to shoot prairie dogs." His first school was a log cabin and boys had to carry knives the first day to kill the pack rats.

Jack Bailey, 88, Montana cowboy
(Original story by Rebecca Colnar)

FROM TOP: *At one year old.*
➤ *Jack Bailey in 2018.*
➤ *Age eight with his range tepee. His first real job paid $1.30 for 13 days of work.*

John L. "Jack" Bailey was born on Dec. 5, 1930, in Montana just after his father, John, passed away from typhoid fever. Jack was to become a fourth-generation rancher in Rosebud County, where his family started ranching in 1883.

Neighborhood Crossroads

© Todd Klassy

Tracks in the snow mark the intersection of Warrick Road and Cow Island Trail, one of the only wagon roads out of the rugged Missouri Breaks to the Montana grasslands. Freighters loaded goods arriving on steamboats from the Upper Missouri River and hauled them on this trail to Fort Benton, sometimes with teams of as many as 12 span of mules or oxen. Faint traces of the original route can still be seen. The rough country south of the Bears Paw Mountains is home to ranchers who may live 30 or more miles down these gravel roads. Signs like this one at rural crossroads make the wide-open spaces feel like a neighborhood.

Pioneers Who Broke the Texas Sod

Nothing was easy for the cowboys in Hardeman County.

By Marge Bennett

Texas is known by some as a God-forsaken place, and by others as heaven on earth. Hardeman County, on the plains of Texas only a short gallop from the Oklahoma border, is the setting for the stories of some of the early settlers, including my grandfather, John Ellington Easley.

Ellington, as he was called, was born in Talladega, Alabama, in 1837, but he and his brothers migrated to East Texas searching for new frontiers and a better way of life. When the Civil War started, he made his way back to Alabama and enlisted in the Army of the Confederacy. After the war, he returned home

more girls were born. The Easleys settled in Big Valley, where they were able to get all the land they wanted and it was cheap virgin scrub brush, windswept and sun-baked.

Farmers would work all year to get their crops ready for market, only to have a hailstorm or tornado destroy their grain. One year, the Easleys paid their taxes with money received for bounty on jackrabbit ears.

night, Grandma would fix a pan of warm water and add a shovelful of ashes to bathe his feet in so that he could walk the next day.

The boys, some of whom were almost grown, helped on the farm and soon Ellington had breeding stock for fine horses, and a good start on raising longhorns. The open range was perfect for cattle and some of the settlers became cattle barons.

The Easleys were not among the wealthy. They raised most of their living, with their own corn and wheat and some cotton. Their cows and hogs furnished milk and meat for the family. They would have to wait for the weather to get cold enough to chill the meat. After butchering a hog, they would trim it and pack it in wooden boxes between layers of salt. They couldn't cure beef that way, and with no refrigeration beef would have to be used immediately. Beef clubs were the answer, with different cuts rotating among the members: steaks this time, a not-so-desirable cut the next, so that everyone had a

LEFT: John Ellington Easley and his bride, Sarah Jane Cliett, true pioneers of Texas before the turn of the 20th century. BELOW: Wedding photo of George and Della Whitton Easley, the writer's parents. Marge is their seventh child.

and married his childhood sweetheart, Sarah Jane Cliett, in 1865. Sick of war, they longed for a new life. They soon joined the ranks of restless citizens traveling west, and since Ellington had spent some time in Texas, he took his new bride and joined a group of friends and relatives moving by oxen-hauled wagons. Their first stop was in Cass County in eastern Texas.

Twenty years later, Ellington and his family moved to Hill County and then to Tarrant County—where they stood on their front porch and waved when the first train came into Fort Worth. Along the way, one girl and eight boys were added to the family. When they finally reached Hardeman County, two

When they decided to make their home base there, it took hard labor and more courage than they had ever expected to break the land and make a home. The town of Chillicothe was just getting started, with business offices, a small grocery store and even the post office housed in makeshift dugouts.

My father told the story of his papa, Ellington, who with oxen and walking plows spent months breaking the sod to make it ready for planting wheat. He wore heavy brogans and the soles were attached with little wooden pegs. Those pegs would work up through the soles and make his feet bleed as he walked and walked behind the plow. At

chance at the best meat. They went to market twice a year for tobacco, sugar, coffee, seeds and medicine.

After a few years, conditions improved and the Easleys built a schoolhouse on one corner of their property. The building was also used as a church and meeting hall, for dinner on the ground, singings, or a dance.

The Easley family in 1892, including widowed sister Annie Harper with her two babies. The sons are all wearing suits, handmade by their mother, Sarah Jane. Left to right, standing: George (the writer's father), Oscar, John Ellington, Elbert, Fayette, Walter and Henry. Sitting: Emily, John Ellington Sr., Janie, Sarah Jane, and Annie with son, Bennie, and daughter, Sammie. BELOW: Grandmother Sarah Jane with her horse Old White Man and granddaughter Marie.

The schoolteacher needed a place to stay so he bedded down with the seven Easley boys in the upstairs of the Easley home. The first school year had a severe winter. This was before horse-and-buggy days, so horse-drawn wagons were used for travel, with hay in the bed of the wagon and quilts to keep the young'uns warm. The wagons would go from farm to farm, picking up children and adding other quilts. My daddy was a trapper and he would check his traps, then walk to school.

The Texas plains were loaded with buffalo bones bleached white by many years in the sun. Hunters had come through in waves shooting the buffalo, skinning them for their hides, then leaving the carcasses to rot. This gave the children a chance to make a little extra money by gathering the bones until they had a wagonload. Then they took them to the railroad and shipped them to where they would be made into fertilizer.

Some years the farms would do well, but

there were lots of years that would bring heartbreak to the county. Farmers would work all year to get their crops ready for market, only to have a hailstorm or tornado destroy their grain. One year, the Easleys paid their taxes with money received for bounty on jackrabbit ears. By 1900, the Easleys had fruit trees and garden produce to help with expenses. They made butter in a 20-gallon churn, then sold and delivered it twice a week to the residents of the nearby

town of Quanah. They sold the butter for 15 cents a pound; corn for 25 cents a bushel; oats, 15 cents; wheat, 25 cents; and eggs, 10 cents a dozen.

My grandfather Ellington worked very hard, but so did my grandmother, who was an amazing woman. She bore 11 children, buried two too soon, and never saw her mother again after leaving Alabama. She carded wool, wove the material, and handmade suits for her big family of boys. The girls learned very early on that they had responsibilities too, and that everyone working together could do great things. In spite of all the hardships, they flourished, and if their Texas wasn't exactly heaven on earth, it was far from God-forsaken. ∎

Marge Bennett sold her first story to RANGE magazine when she was in her nineties, then she taught writing to other seniors in Reno, Nevada.

CONFESSIONS OF RED MEAT SURVIVORS

Some of ranching's old-timers admit to ignoring the problems of cholesterol and other unnamed and often unsubstantiated handicaps. They believe that red meat is good, which is proven here, simply by age and attitude.

Maxine Sweeney, 93

An excellent life.

In 1916, the year that Maxine Maupin was born, windshield wipers, bobby pins and guide dogs were introduced, the first PGA Tournament was held, and the first Norman Rockwell painting appeared on the cover of the *Saturday Evening Post*.

Born in a tent on Friday the 13th of June in Powers, Ore., to Charles and Emma (Reeder) Maupin, the eighth of nine children, Maxine has and continues to live a healthy and adventurous life. Her childhood was spent by the Illinois River in Oregon on a 160-acre homestead where her family lived in a cabin built by her father with logs he hewed from local trees.

"I didn't realize the times were tough," Maxine says, "as I never remembered being cold or hungry." There were the deaths of two siblings from pneumonia and from infection and the family also lost everything they owned when their house burned to the ground. Money was scarce and Maxine's mother made most of the children's clothes. "I had to be careful when I bent over," Maxine says, "so as not to reveal the 'Best Whole Wheat Flour' printed across my rump." The rest of their clothes came from the "Monkey Ward" catalog.

There were no roads in the area and the mail boat delivered letters and groceries to Riley's store at Agness twice a week. "I traveled by boat and train to get to school, not ever worrying about the dangerous high water in early spring when crossing the river, nor the bindle stiffs that rode the train and walked the tracks."

The family moved to Gold Beach, Ore., when Maxine was nine. She says they felt like "hicks from the sticks." Her flour-sack bloomers and long hair did not make a fashion statement. From there they moved to Reedsport.

Maxine was a straight-A student and an athlete, excelling in basketball and swimming. "I graduated from high school in 1933, when my family moved to Fort Bidwell, Calif., due to my father's health problems." Maxine met Paul Sweeney at a Halloween dance, and they were

> ## "I had to be careful when I bent over, so as not to reveal the 'Best Whole Wheat Flour' printed across my rump."

married a year later in Reno, Nev.

Paul and Maxine moved to Golconda, Nev., to run the Diamond S Ranch where Lester was born. They continued to work for local ranchers and later in the mine at Midas. Then they moved to the Squaw Valley Ranch, owned by the Ellison Ranching Company, which Paul ran 21 years. They raised four children: Pat, Gene and Lynn joined Lester and the Sweeney family.

"I was fill-in cook for the ranch crew, and transported children to school in Midas, sometimes by bobsled with a team of horses named Laddie and Dick when the snow was deep." She measured the snow depth on skis for the state of Nevada. She has many interesting stories of the years she acted as justice of the peace and coroner in Midas, trapping

PHOTOS COURTESY MAXINE SWEENEY

FROM TOP: Maxine in center with brother and sisters in the Illinois River.
➤ *Paul and Maxine's wedding, 1934.*
➤ *Maxine's 90th birthday and she's still rattling pots.*
➤ *Maxine, 3, on a mule at the homestead.*

coyotes and bobcats from her Model T or horseback, and riding with Paul when he was out with the buckaroo wagon.

Paul and Maxine then moved to Winnemucca, Nev., where he worked as a government trapper. "We built a bar and restaurant in Orovada," Maxine says, "and Paul continued to trap while I ran the bar. After we sold the place, I drove a school bus."

Then they moved to Emmett, Idaho. Paul passed away in 2001 at the age of 92 and Maxine now lives on an acre. She gardens, cans fruit and vegetables, and rides her exercise bike a few miles a day. She travels to see her great-grandchildren participate in sports, rodeos and team roping, and takes in a few spring brandings. "Every morning I take a shot of vodka with raisins to prevent arthritis." She eats red meat almost every day with fruits and vegetables, usually from her own garden.

"Things that irritate me the most," says Maxine, "are my kids telling me what to do, overzealous environmentalists, and doctors."—*Gene Gabica*

Raymond Milburn Green, 103

Slow down!

Raymond Milburn Green was born in Bentonville, Ark., to Mary McBride and Clyde Milburn Green on Oct. 1, 1912. "My grandparents had a variety store there. I often wondered if Sam Walton bought their store when he started Walmart."

His family moved 32 times in 21 years. Raymond went to school in Topeka, Kansas, and worked in stores before and after school. "After the eighth grade my folks thought I'd had enough schooling, so I was on my own." At 12, he worked as a soda jerk. "By the time I was 15, my $7 a week working at a grocery store supported my family as my father had lost his job."

They moved to Montrose, Colo., in 1920. Ray found work on a ranch even though cows scared him. He is a small man and he had to work twice as hard to pitch bundles onto a wagon. He later worked for Dick and Bertha Lloyd in Mesa, where he broke lots of workhorses. His father died at 48 of tick fever.

Ray married Mary Ann Heaton on Oct. 1, 1938, in Montrose and they had three daughters—Rosalee Milburn Dean, Maralee Stiegel, and Carolyn Scott. He managed to save $200 for a down payment on the $850 it cost him to buy nine acres and a shack near Fruita. He also bought a milk cow. "The shack was full of bedbugs so my family stayed elsewhere until I could clean up."

Ray helped WPA and CCC boys build the road in the Colorado National Monument. He was timekeeper, surveyor, secretary (no one else could type), and general flunky. "I got paid $44 a month and drew the first map of the monument that included Rim Rock Drive. They used it on their brochures."

Ray worked for the Bureau of Land Man-

> **"When I was a kid my folks were always gone when I came home from school," Ray says. "I had to do the cooking and cleaning. I didn't mind the work but hated being alone."**

agement as a mapmaker and surveyor. After Pearl Harbor he was drafted, but the Army didn't want him because his feet were too small. During the war he helped build air bases for the U.S. Army Corps of Engineers in Missouri, Iowa, Wyoming and South Dakota. He helped improve the Lowry Air Force Base in Colorado, and says, "I liked contributing to the war effort."

After the war he helped design big irrigation projects for the Bureau of Reclamation, including Glen Canyon Dam in Arizona. He worked on a seven-by-12-foot map that was used by Congress on its vote to create Lake Powell on the Colorado River. He also worked on the Manhattan Project for the Atomic Energy Commission. He operated his own printing shop for five years and manned presses in the regional shop for the Bureau of Reclamation. He even worked to relocate London Bridge in Lake Havasu, Ariz.

After 53 years away, Ray and Mary Ann moved back to Fruita. Mary Ann passed away in 2008 after almost 70 years of marriage. Ray never wanted Mary Ann to work. "When I was a kid my folks were always gone when I came home from school," he says. "I had to do the cooking and cleaning. I didn't mind the work but hated being alone."

Ray has been a member of the Masons since 1947. He misses Mary Ann, but enjoys their five grandchildren, eight great-grandchildren, and three great-great-grandchildren. The Mesa County Cattlewomen chose him to be Father of the Year in 2015.

At 98, Ray bought a house and got a 30-year loan on it. After he renewed his driver's license recently, he bought a new car. He drives down to Judy's Family Restaurant every day to eat and visit old friends. On one of those trips, he was stopped for speeding. The cop said: "I have never given anyone over 100 a ticket and I won't give you a ticket now. Just keep it down."—*Patty Vaughn Miller, 85, writer, artist and "wornout cowgirl"*

PHOTOS COURTESY GREEN FAMILY

CLOCKWISE FROM TOP: Ray Green helped WPA and CCC as timekeeper, surveyor, secretary (no one else could type), and general flunky. "I got paid $44 a month." ➤ *Ray and Mary Ann, ca. 2005.* ➤ *The couple while dating in 1938.* ➤ *Mesa County Cattlewomen named Ray Father of the Year on Father's Day, June 20, 2015. He's celebrating with members, from left, Jill Berthod, Patty Miller, Debbie Albertson, and president Verla Rossi.*

Jessie Kist Hammond, 100

Walking with the cows.

"Motion is lotion," says Jessie Hammond (née Kist) of Scott Valley in Northern California. "The secret to a long life is to do a lot of walking and to eat your vegetables."

Whatever her trick is, it seems to be working. Jessie turned 100 on July 7. At her birthday party in the Etna city park, several hundred well-wishers gathered. Most of them called her "Aunt Jessie" or "Grandma Jessie," whether they were blood relatives or not.

Jessie is famous around Scott Valley for chopping Marlahan Mustard (Dyer's Woad) and her own firewood; taking cows to the mountains afoot; out-farming the hardest-working men; and wearing a broad-brimmed hat, sweatshirt, and smile that won't quit. She still loves walking through her cows, raising her garden and living in the drafty 1892 ranch house on the place she and her late husband, Carl, bought in 1945. They raised Angus cattle and four kids on the ranch, located just outside Etna.

"My husband said black cattle and chopped hay was the only way to go," Jessie says. "So that's what we did."

Though she took to ranch life, she started out a "town kid," born and raised with four sisters in the town of Etna—population today, 737. Her father, also born in Etna, owned a flour mill. Jessie fished with her dad and enjoyed lots of sports. She played town team softball until she was 60, loved tennis, and was a faithful fan at all the high-school home games until well into her 90s.

Jessie married Carl Hammond after her sister, Anna, married his brother, Dwight. Jessie's energy and can-do attitude made her a perfect match for Carl. "He never quit," she remembers. "We got married, never

COURTESY GERALDEAN CHRISTOPHER

Clockwise from top: Anna, Jessie, Emma, Deanist and Betty Kist. ➤ *Carl and Jessie on their wedding day, Oct. 23, 1937, with preacher and friend.* ➤ *Jessie (in the middle) with tomboy girlfriends.* ➤ *Jessie (wearing a favorite floppy hat) and her great-grandniece, Kiely, pose in front of Jessie's house.*

PHOTOS COURTESY HAMMOND FAMILY

> **"Everyone said working alongside Carl would kill me. He never quit. Now they're all gone and I'm still here. I liked out there working with him."**

went on a honeymoon. He went to work fall-bucking trees."

Carl was an unmatched logger. He won many titles for cutting timber with a crosscut saw. When all the other loggers would go home, he'd keep working, using a crosscut saw rigged up for one man.

Jessie joined him in the woods on weekends as he worked to earn the down payment for their ranch. They bought the Pete Smith place in 1945 and eventually developed a prize-winning line of registered Black Angus.

Never fond of riding horses, Jessie has walked hundreds of miles over her lifetime. Well into her 80s, she trailed the cattle by foot to their summer range on U.S. Forest Service

ground—a 40-mile round-trip. But she *would* ride the tractor Carl bought her because she liked farming.

"My tractor had 'Jessie' on it and his had 'Carl' on it," she says. "And we got two plows so when we did a field it went twice as fast. It worked good for us."

She still likes driving tractors even though the tip is missing from her right ring finger. "I was cutting hay, and it got plugged up. I didn't shut the stupid mower off. I reached in and chop." She showed it to Carl, who made her go to the doctor. "Otherwise I was gonna come in, wrap it up and go on working."

Carl passed away in 1996, leaving Jessie to run the ranch with the help of her children and numerous grandchildren. Now her two boys own the ranch but Jessie still has cows.

"Everyone said working alongside Carl would kill me. Now they're all gone and I'm still here. I liked out there working with him," she says, looking out the window. "Yep, away we went."—*Theodora Johnson*

End of the Day with Grandpa

© Connie Thompson

Kenny Calloway and his granddaughter Devan Bauer were close, often riding together to check cattle on the Thompson Ranch in East Texas, where Kenny was foreman. In this photo Kenny is about 50 and Devan is seven years old.

Devan idolized her grandfather, loved to spend time with him, and soaked in all the knowledge he had to offer. Kenny loved being with Devan and admired her enthusiasm and intelligence.

Sadly, Kenny died of Lou Gehrig's disease in 2020. Devan misses her grandfather greatly, but put the lessons she learned from him about dedication and determination to good use by earning a degree in nursing.

Don McPherson, 84
Packer and warden.

Anglers have enjoyed fabulous fishing in Idaho's Selway Bitterroot Wilderness for many years, but very few have any idea how the fish got there in the first place. Don McPherson can tell you.

Don was born in the family home near Kooskia, Idaho, on July 3, 1926. His grandfather was a pioneer ranger who supervised and mapped much of the present Nez Perce National Forest. His father packed and guided for years in that forest and also in the Selway Wilderness. Don learned the packing trade from him. He chuckles when he says, "I went with Dad until I knew better."

Don left high school in 1943 to join the Navy and served in the South Pacific on the USS Whitney, a destroyer tender. He returned home after the war to graduate from Kooskia High School and go to work for the U.S. Forest Service. He packed mules and horses into the Lochsa and Selway drainages for a year before joining

ings. The fresh water cooled the fish and provided additional oxygen as well. The fingerlings were released very slowly into lakes and creek water was often added to the release bucket to prevent icy cold lake water from shocking the fish. Some 60 years later, fly fishermen still enjoy success

PHOTOS COURTESY DON MCPHERSON

in Glacier National Park. Lorraine often rode with him as he delivered. "One time, I packed two cases of dynamite around a steep cliff and then through a tunnel. My mules balked in the tunnel and finally emerged on the other side to the sound of loud yodeling." Two Norwegian lads who

> "One time, I packed two cases of dynamite around a steep cliff and then through a tunnel. My mules balked in the tunnel and finally emerged on the other side to the sound of loud yodeling." Two Norwegian lads who had been hired to shovel snow and ice from the trails were yodeling at the top of their voices. Don hollered: "Stop that yodeling or I'll have to shoot you both. You are scaring my mules!"

Idaho Fish & Game in 1947. Over the next two years, Don packed 40 tons of salt into the mountains during the summer months for big game animals. He also stocked 15 mountain lakes and seven streams with cutthroat, rainbow and brook trout.

Most of the lakes and streams were 15 to 20 miles up and over steep ridges on dangerous trails. Fingerling trout were packed in by mule train. Each mule carried two 10-gallon milk cans loaded with water and about one-and-a-half pounds of fish. Burlap covers allowed water to slosh out and make room for more at creek cross-

FROM TOP: Don rides Avalanche while packing in Glacier National Park, 1952. ➤ Don leads a string of 10 pack mules over Big Fog Saddle. Mules are loaded with trout fingerlings for stocking lakes in the high country of Idaho's Selway-Bitterroot Wilderness. ➤ Lorraine and Don in 2010.

on these high mountain waters.

In 1949, Don married his high school sweetheart, Lorraine Yenney. He moved to the National Park Service and trained and packed mules

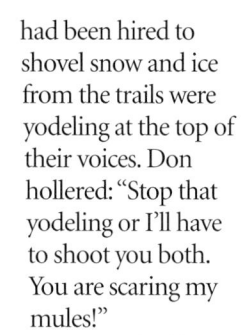

had been hired to shovel snow and ice from the trails were yodeling at the top of their voices. Don hollered: "Stop that yodeling or I'll have to shoot you both. You are scaring my mules!"

Don retired in 1981. He still has a saddle horse and four mules but he doesn't ride much anymore. He looks fit, gets around well, and tells some great stories about his days in the saddle.—*Phil Lamm*

Beth Smith Aycock, 95

`Everybody's outdoor sport was cattle rustling.`

Beth Smith has been a cowboy since birth. At 95, she likes to drink a little tequila, eat beefsteak and frijoles, and dreams of riding the range again. "I'm like a wild steer all penned up," she says of her pleasant retirement apartment in Green Valley, Ariz. "I never have lived in town."

Beth was born on a ranch in New Mexico's Guadalupe Mountains. Her father, Red Howell, ran up to 3,000 head of cattle over brushy, alkali country that she calls the roughest in the world. He was a "rider, rancher, roper and rounder" who'd go on benders, leaving the ranch to her mother, Nell Choate, a former schoolteacher.

Beth knew all about keeping up with boys. She was the lone female among three siblings and 28 cousins. "I had to learn to ride, shoot, rope, everything the boys did," she says. "I couldn't cuss or chew tobacco, but I wanted to."

At 17, Beth entered the University of Arizona, but left when her father went bankrupt. "We lived so far out it was impossible to go to school," she says. "I didn't really want to because I had to make a hand."

Growing up, Beth lived on five ranches in Arizona and New Mexico and another nine after marrying Sam Breckenridge Smith in 1939. All were remote. When she brought her only child, Rick, home from the hospital in 1945, she had to ride eight miles on horseback to reach the A7 Ranch east of Tucson with her infant resting on a pillow on her saddle.

The scarcity of water was constant. She remembers Sam rigging up a sled, putting a 50-gallon water barrel on it, and hauling water six miles from their windmill down a rocky canyon. "I used that water to bathe Rick, wash his clothes, mop the kitchen and water our trees," says Beth. "Sam gave me a glass of water in the morning and said, 'You can either take a bath in it or drink it.'"

On the Lazy V Ranch, Beth and Sam partnered with Bill Veeck, then owner of the Cleveland Indians. Between 1950 and 1970, the Smiths and Veeck partnered on the Deep Creek Ranch, 80 miles north of Silver City,

> **The scarcity of water was constant. Sam rigged up a sled, put a 50-gallon water barrel on it, and hauled water six miles from their windmill down a rocky canyon. "I used that water to bathe Rick, wash his clothes, mop the kitchen and water our trees. Sam gave me a glass of water in the morning and said, 'You can either take a bath in it or drink it.'"**

N.M. The land reached 9,000 feet, measured 35 miles long and encompassed 96 sections. "It was in Catron County, the biggest in New Mexico and the last county courthouse to have electricity in the U.S.," says Beth. "It was a hundred years behind the times and everybody's favorite outdoor sport was cattle rustling."

Beth recalls the day Sam failed to come home. She found his riderless horse at dusk near their big pasture and followed its path three miles over some mesas to Deep Creek. "There was old Sam," Beth says. "He'd done something very foolish."

He had encountered a bear cub crying in a tree and thought something had happened to its mother. To save the cub, he ground-tied his horse and climbed the tree just as the mother came back, boogered his horse, and left Sam afoot.

"Those were the happiest years of my life," says Beth of Deep Creek. "I got to ride every day and watch my son grow up."

Sam died after 37 years of marriage. In 1982, Beth was running the Sonoita Fairgrounds when she married Bill Aycock, an engineer for General Motors. "He married me because I had a new pickup and a five-horse slant trailer," she jokes.

In addition to cowboying, Beth has published seven books, including "A Red Howell Fit" about her father. She plans to start on another book about her mother's family coming from Texas to Arizona in the 1880s. That assumes no more setbacks like the broken hip she suffered last May.

"I've been thrown off a horse, run over by cattle, stepped on, kicked and everything else in the world," Beth says. "And I come into a nursing home and fall and break my hip. Don't worry. I can still kick, but not very high."—*Leo W. Banks*

PHOTOS COURTESY BETH AYCOCK

CLOCKWISE FROM TOP: Beth today in Green Valley, Ariz. ➤ *Branding at the A7 Ranch, ca. 1944. Beth is at right, holding the calf's heels.* ➤ *Gabriel and Matilda Choate, Beth's grandparents, ca. 1900 on their way from Turkey Creek to buy supplies in Douglas.* ➤ *Beth with son, Rick, and husband, Sam, at Deep Creek Ranch in 1955, "the happiest time of my life."* ➤ *Beth rolls a Bull Durham smoke at Kenyon Ranch in 1940.* ➤ *The start of roundup at the A7. Beth is #2 on her best horse, Deadeye.*

Long Dust on the Desert

Edward M. Hanks, a buckaroo's tale, with excerpts from his book.
By Carolyn Dufurrena. Photos courtesy Ed Hanks Jr.

Edward M. Hanks left his home ranch near Fort Bidwell, Calif., in spring 1900. He was 18, one of 11 children, off to seek his fortune in Nevada with his cousin Frank Phillips. He ran mustangs and buckarooed on ranches in northern Nevada and Oregon in the first half of the last century, traveling back to Modoc County in between jobs.

Ed's travels brought him to the Hunter and Banks Ranch in Elko County in 1912. There were no fences. Thousands of cattle roamed on the ranges. He rode with the crew of some 25 buckaroos. Several local ladies, notable for their woolly chaps, white blouses and ties, helped ride as well. One of them, 15-year-old Ella Eddy, whose mother was the local schoolteacher hired by Hunter and Banks, entranced Ed since 1916.

Ed and Ella buckarooed together for several years and finally married in 1923. She was 22; he was 40. Their middle son, Ed Junior, now 92, remembers what it was like living the buckaroo life with his parents. "I was two years old when I started riding, in front of Mom's saddle. By the time I was six or seven, I rode a lot with the cowboy crew."

Ed Junior remembers one trip from the Horseshoe Ranch trailing 130 bulls 100 miles north to Lone Mountain. "It took three days, and there was only one bullfight. The first night we

Ella and piglet, ca. 1922-23, during the time that Ella was teaching school on the Hunter and Banks Ranch, shortly before she and Ed married.

Chow time. This is probably at the Stampede Ranch, north of Red House. Thousands of cattle grazed in common across hundreds of square miles of unfenced land. Cowboy crews of 20 or more rode to gather the herds and bring them to the railroad for shipping. The Hunter and Banks outfit included half a dozen smaller homesteads, which the company absorbed over time, including the Huntsman, Hadley, Horseshoe and Green Cabin.

Cowboys relax on a pile of tents and bedrolls while the wagon goes in search of firewood. Camp gear is in boxes at right.

Skylined, the cowboy crew pauses for a photo. We suspect the riderless horse, which appears in most images, belongs to the photographer. In his book, "A Long Dust on the Desert," Ed Hanks wrote: "We had over 100 horses in the saddle horse herd. We had two men to wrangle them. In rounding up the horses to ship, a couple of us would go to the ridges to start the bands of horses down the trails. Several more men would go down near where we wanted to bunch the herds. This was very exciting and riding was fast. Dust clouds began to rise in all directions, and in a little while, the horses came in with streaming manes and tails, all kinds of horses, with a few antelope and deer among them for variety. They could be seen running near the rodeo ground, where riders would herd them into one bunch. It wasn't always easy to slow these beautiful, high-spirited animals down. The deer and antelope usually found their way out before the horses were bunched. When we got them all together we drove them to corrals, where we branded the colts and worked out the shippers to hold in a field until we got enough to make a drive to the railroad. We had to neck all the young colts to older saddle horses to hold them, as our wranglers would put them out on the hills where the grass was good at night. In June, the longest days of the year, we would be necking those horses together after dark."

Ella Hanks rode and roped with the cowboy crew. Her mother, Florence Eddy, had moved back to Elko from California and was hired by Hunter and Banks to teach school on the ranch after her miner husband died of consumption. Ella met buckaroo Ed Hanks when she was about 15 and they rode together for several years, until she "came of age." They were married in 1923. "After Dad and Mom had us three boys," says Ed Junior, "my grandmother quit teaching and joined our family so Mom could buckaroo."

Mustang trap west of Simon's Place on Maggie Creek Ranch. One of Ed's first jobs in Nevada was running mustangs for the Taylor and Edson's Double Square Ranch north of Winnemucca. "They didn't know how many horses they had," Ed wrote. "They didn't have any way to count them, but they would gather at least a thousand to sell each fall, four years in age and up."

Ed Senior with Mag and Mollie. Ed ran the Horseshoe Ranch near Beowawe in the 1930s with a remuda of about 100 head of saddle and workhorses. He bought them "mostly from ranches north of Elko," where they ran outside, often joining bands of mustangs.

Five boys on Smokey. From front to back: Roland Hanks (the eldest of the three); Gordon Smailes (a friend from Elko) and his brother, Jack; Ed Hanks Junior, the littlest one, and his younger brother, George. Ed Senior found Smokey near Gravelly Ford on the Horseshoe Ranch near Beowawe, suffering from a badly infected wire cut across his chest. He bought him for $50 from a local Shoshone boy, Bill Buffalo, before Bill left for the Stewart Indian School in Carson City. Smokey weighed about 1,200 pounds and had a reputation as a bucker, but Ed doctored him back to health and Smokey became one of the family. Ed wrote: "He was a wonderful horse, good-natured and gentle. He liked water and could swim like a duck."

kept those bulls in a guard corral, and the next morning, though there was no break in the fence, we found one bull on the outside. We finally figured out that two of the others had hooked him with their horns and thrown him over the fence. We found the place where he landed on the outside."

Ella was along on that trip too, cooking on the wagon. Her husband was approaching 60; Ed Junior was 10. (Jon Griggs, current manager of what is now Maggie Creek Ranch, says, "Those would have been three pretty long days.")

The family buckarooed together for the next several years, moving from the Horseshoe Ranch to the range on Maggie Creek in the summers, living in tents. Ed Senior managed the outfit until 1943, when he retired from the back of a horse. He put up hay for the outfit for quite a few years after that but went to work supporting the war effort by guarding the Carlin tunnels for the Union Pacific Railroad—which ran through the headquarters of Hunter and Banks Ranch.

He wrote about the Horseshoe Ranch and his time at Hunter and Banks in his 1967 memoir, "A Long Dust on the Desert."

Ed lived to be 92 and was inducted into the Buckaroo Hall of Fame in Winnemucca, Nev., in 2011. ∎

Carolyn Dufurrena lives on an old Miller and Lux ranch in northwest Nevada. Ed Hanks stayed there on the way to his first job in Nevada. Ed's book can be purchased through the Winnemucca Buckaroo Hall of Fame Facebook page.

Ella Banks, second rider from the left, heels with the girls while the buckaroos work the ground in 1916. Ed, in a white shirt, kneels behind the calf at left.

Three ladies in woollies with Ella Hanks on the right.
"After Al Phillips was married, his wife, Bea, would ride with us sometimes," Ed wrote. "She liked to have the Webber girls—Belle, Clara and Mabel—come out too. These girls were really good riders, as their father had trained them to help with his cattle. So they were riding with the cowboys at Stampede. Willie Davis was the cowboss and it was September 12th, and Bea and I had a birthday. Willie said to me, 'If you can catch some of these chickens around here, you can have a chicken dinner.' I hunted two dry sticks about four feet long. I would twirl them through the air at some likely looking young roosters, and it wasn't long before I had some nice fryers ready to pick. Later on that day the women fried them and we had a fine chicken dinner. Mabel said, 'Ed, if I was raising chickens, I wouldn't want you to come around there.' They all laughed at that."

Hand in Hand

© *Cynthia Baldauf*

Grandpa Harold Peterson (left) and neighbor Jon King hold Malcolm Peterson steady as he learns to walk in his new cowboy boots. It won't be long before he joins in with his own horse and helps drive the herd 18 miles to summer range near Jackson, Montana.

Rudy Hergenrider, 92

Lots of horses.

Rudy Hergenrider was born in 1926 in the house he still lives in today. His German parents came over from Russia and settled near Belfry, Mont., about 70 miles south of Billings. When Rudy was a boy he walked two and a half miles to and from school. He also dug sugar beets by hand—a long, hard job during harvest season.

One of Rudy's best memories is gathering horses on the open range. Whether they were draft or saddle horses, farmers and ranchers turned their horses out during certain months of the year so they could graze freely. Horses were needed for spring plowing, summer haying, and fall cattle gathering. Between their working seasons, horses were turned out on open range in Elk Basin and the Cottonwood area—108 sections, or about 70,000 acres east and south of the Hergenrider farm.

"When you needed horses, you'd take the horse you had in the corral and ride out to gather the herd," Rudy remembers. Rudy's favorite horse was Duke, purchased from Barry O'Leary out by the Indian caves. Nobody but Rudy could ride him.

"Sometimes as a kid I would ride for a few hours, then lie down and sleep for a while, then get up and go again. I got kicked by a stud horse once and had to walk back to the farm."

He adds: "One time we were in our buggy going to church and my dad saw our neighbors using our team. 'Hey, those are our horses,' he said. The collars and harnesses did not fit those horses right so we had to get them back from our neighbors. That happened from time to time—someone would end up with your horses."

The Hergenrider family had cattle, as well as a few sheep from time to time, and hogs. "We had a large family," Rudy says, "and we'd make a circuit with all the relatives in the fall, butchering a steer at each place and a couple of hogs so we all had some beef and pork for the winter."

When Rudy enlisted in the Army in 1944 he was stationed in Alabama, but his desire was to be overseas. He spent 18 months in the Philippines, where he served

CLOCKWISE FROM TOP: The Hergenrider family. Back row: Dave, John, Alfred and Rudy. Middle row: Emma and Olga. Front row: Anna, Dave (father), Lydia and Sophia (mother). ➤ *Rudy and brother Alfred in military uniforms, 1944.* ➤ *Rudy in 2014.* ➤ *Irrigating pasture in 2010.* ➤ *With an unidentified friend.*

as a guard at the presidential palace during the war crimes trials in Manila. "I was there when General Tomoyuki Yamashita, the 'Tiger of Malaya,' and General 'Death-March' Masaharu Homma were put on trial." They were convicted and hanged in 1946. Rudy was on the team that transcribed the trials. He brought a copy of the transcripts back home, but they were unfortunately destroyed in a flood.

After the war, Rudy went right back to farming. "My friends and neighbors who had stayed behind to farm had already transitioned to tractors," he says. "I had to make do with horsepower until I was able to work up to a tractor."

Rudy's first automated farm equipment was a B series John Deere tractor, the first in the line of green farm equipment for the Hergenriders. "When I came home from the war, I owned 80 acres, 40 irrigat-

Between their working seasons, horses were turned out on 70,000 acres of open range. "When you needed horses, you took the one in the corral and rode out and gathered the herd. Sometimes as a kid I would ride for a few hours, then lie down and sleep for a while, then get up and go again."

ed. Now I have 345 acres irrigated and 655 acres of range ground. I steadily grew my cowherd, as well. Like everyone else, we had Herefords and then transitioned to Black Angus in the 1970s."

He married Mary Wennemar in 1951 and they had four children. Mary passed away in 1991, and Rudy married Clara Harper, who passed away in 2013.

The former soldier was recognized in 2013 and invited for the Big Sky Honor Flight to Washington, D.C. It was created solely to honor America's veterans.

—*Rebecca Colnar*

Gene Aguirre, 100

It was pretty tricky
a time or two.

Gene Aguirre was born Sept. 10, 1909, in the Arizona Territory. As a boy, he listened to stories told by his dad and uncles about the real Old West and today Gene is still recounting stories about ranching, Indians, freighting, old vaqueros and the history of his pioneer family with a strong and clear mind.

The Aguirre families emigrated from Spain to Mexico in the late 1600s. They played a part in the settlement of Mexico and soon were among the leading families there. By the mid-1800s, however, political ties were strained in Mexico when Gene's great-grandfather backed the wrong politician who wound up losing an election. Then, for "health reasons" (fear of being shot), the

his great-uncle, Epifanio Aguirre, who made headlines by being credited with saving the stage that ran from Socorro, N.M., to El Paso, Texas, back in 1864.

"Uncle Epifanio was traveling with his wife, their two small children, two servants and a saddle horse in a big four-wheeled

> ## "When attacked, Uncle Epifanio mounted the saddle horse and galloped out in front with a pistol in each hand and the bridle reins in his teeth. He would empty his pistols and clear a path for the stage and then gallop back to his coach where his wife would hand him two more loaded pistols!"

clear a path for the stage and then gallop back to his coach where his wife would hand him two more loaded pistols! It was said that Uncle Epifanio showed courage above and beyond what most had ever seen before. After several miles of a running battle, a small village came into view and it was only then that the Apaches finally quit the attack."

Gene grew up living a dream on his father's large ranches near Red Rock. They ran up to 10,000 head of cattle in their heyday and young Gene grew up working with some of the best cowboys in the Southwest. They were actually vaqueros who had been practicing the art of handling cattle for generations. "I grew up learning the art of cowboying from the best vaqueros in the land and learning about being a cattleman from my own family."

Gene's Uncle Pedro is credited with developing the famed Buenos Aires Ranch along the Arizona/Mexico border. This ranch was a stage stop in 1864, went on to run 15,000 head of sheep and cattle during Pedro's tenure, and now it's a wildlife preserve.

In the late 1940s an event happened that moved Gene into another exciting direction. The dreaded hoof and mouth disease was discovered in Mexico and men from the southwestern United States who could speak Spanish and had a ranching background were asked to help fight the disease.

"I worked as a livestock inspector, a supervisor of livestock inspectors, and as an appraiser of livestock," Gene says. "I saw things that you would not believe today. I forded swollen rivers during flood season on a mule, crossed high mountain passes where the trails were narrow, slippery and treacherous, and even faced down armed men while standing up for my job and what was right. It was pretty tricky down there a time or two!"

When asked about the secret of his longevity, Gene says: "Well, I eat meat, potatoes and beans on a regular basis and have a glass of red wine now and then. But most of all, you have to enjoy life. You can't go around worried and stressed out all of the time. You have to have fun. You need to smile."

—*Jim Olson*

Don Yjinio Aguirre 1889
MULE TRAIN HAULING COKE from WILCOX
To Globe (PICTURE taken in WILCOX)

CLOCKWISE FROM TOP: Gene's grandfather Yjinio Aguirre and his wagon train in 1889, hauling coke from Wilcox to Globe, Ariz. ➤ Gene at 100. ➤ Aguirre Ranch, ca. 1928. From left: vaquero Jesus Mendez, Enrique "Henry" Aguirre (Gene's brother), and Gene. ➤ Young Gene in 1928.

1928

family patriarch decided to move his clan north.

"My Aguirre ancestors were some of the first to haul freight across the Santa Fe Trail from Missouri to New Mexico," says Gene. "And the Aguirres were also some of the first ranchers in the American Southwest. They helped to settle and develop this country through mining, freighting, trading and ranching during the 1800s, just as they had done in Mexico during the 1700s. Read the history books and you'll find information on my pioneer family."

One story that Gene likes to tell is about

> ## "I grew up learning the art of cowboying from the best vaqueros in the land and learning about being a cattleman from my own family."

ambulance. They were following the stage when it was jumped by a band of Apaches.

"When attacked, Uncle Epifanio mounted the saddle horse and galloped out in front with a pistol in each hand and the bridle reins in his teeth. He would empty his pistols and

Giles Milton Lee, 94

Roping and rodeoing.

Giles Milton Lee, the youngest of four children, was born Dec. 10, 1922, in Midland, Texas, to Richard David Lee and Sarah Viola Forrester Lee. He was three years old when he and his family moved from Midland to Lea County in New Mexico.

Dick Lee, Giles' father, partnered with Midland ranchers Clarence and John Scharbauer of Scharbauer Cattle Company to carve out a new life on the

Swamp Angel ranch. Dick trailed 1,500 head of heifer yearlings from Midland across the Llano Estacado to the new ranch near Buckeye. "Bill Kelton, grandfather of Merle and author Elmer, drove the next big bunch of cattle to Swamp Angel," Giles says. "Scabies broke out among the neighboring ranches and cattle had to be dipped for 31 days."

Giles remembers seeing his first rodeo. "I was seven or eight and cars were parked in a circle to define an arena in a pasture west of Lovington." He was hooked on rodeo, and recalls winning "a pair of socks and a tie" in a roping competition. Giles graduated from Lovington High School in 1941 and headed to New Mexico A&M to try his hand at college rodeo competition.

Giles and 13 other rodeo team members organized a rodeo club. He did well, winning

the Wild Cow Milking with Bill Spires at the Eastern New Mexico College rodeo, the All-Around at West Texas State and at NM A&M. In 1942, after a year of college, Giles joined the military. He spent his war in the Pacific on B-17 aircraft. After his discharge in 1946, Giles returned to NMSU (formerly NM A&M) to finish school. "I tried and failed to pass chem-

Clockwise from above: New Mexico State University Rodeo Club, 1941. Standing, from left: Bud Prather, Giles Lee, Jim Rush, Elroy Fort and Smokey Nunn. Sitting, from left: Bob White, Bob Malcom, Pat Patterson and Bill Maxwell. ▶ Joie and Giles on their 60th wedding anniversary in 2007. ▶ Roping in 1989. ▶ Giles in Lovington, N.M., spring 1941.

istry three times! We had started the rodeo team before the war and we reactivated after we all came back to college."

In his greatest achievement, Giles and his team paved the way for new rodeo cowboys and cowgirls to obtain degrees while pursuing their rodeo dreams. Giles kept in touch with the original team members and 60 years later, in 2002, he and four surviving members were honored at New Mexico State

Giles and 13 other rodeo team members organized a rodeo club and paved the way for new rodeo cowboys and cowgirls to obtain degrees while pursuing their rodeo dreams.

University for their pioneering efforts.

He and his wife, Joie, spent 70 years at Swamp Angel, where they raised three daughters, along with Hereford and later Red Angus cattle.—*Sally Anderson Bland*

Crooked Creek Rodear

© Madeleine Graham Blake

At Summit Springs in southeastern Oregon's Sheepshead Mountains, Reuben and Grace Stoddart take a break from branding for the Crooked Creek Ranch and play outside, using their budding muscles for the hard work that most likely is their future. Sixteen miles up a well-traveled dirt road from the home ranch, a stock pond, holding pen and cookhouse make up the Stoddart camp. Cowboys, moms and kids leave this camp to brand calves whose wild mothers spend their lives on open range. Tomorrow they will ride out, Reuben with his mother and Grace on her own horse, to work the cattle that have been grazing free on the vast holdings the Stoddarts lease. Holding rodear, they and a handful of other cowhands will contain the herd while the calves are doctored and branded.

Gene Etchart, 97

Seventy thousand meals.

"My dad, John Etchart, at age 20," Gene says. "came from the Basque Pyrenees to run sheep in partnership with cousins." They herded bands from Montana to Southern California from 1900 to about 1911. The next year, John returned to his native France and married Catherine Urquilux. The newlyweds settled on Montana's Missouri River grasslands in 1912.

Gene, his sister, and three brothers were raised on that remote ranch, which grew to 250,000 acres, running 30,000 sheep together with Hereford cattle. "Our introduction to school included an admonition that our parents quit speaking Basque, as our English was very poor. Mom never spoke Basque in the house again."

They grew up involved in the ranch. "As a 10-year-old, I was part of the crew that drove 25 teams putting up hay. By my

PHOTOS COURTESY ETCHART FAMILY

early teens, I accompanied railcars of cattle to Chicago markets. The trips were scheduled to fit Notre Dame football games. At 97, the Fighting Irish are still my favorite."

Growing up, airplanes fascinated Gene. His mom's brother had died as a military pilot, making his parents reluctant about his interest. "In spite of that, I became a flight instructor and owned an airplane by age 20." His parents gradually came to appreciate that "high-loping buckskin's" ability to help on the ranch. "Dad's regular journey in a heavily loaded buckboard to scattered sheep camps would take a week or more. One day, I gave him an aerial tour of all those camps, landing at some. We were back at headquarters by lunchtime."

Convenience, and coyote control, soon made the airplane essential. That flying experience led Gene to buy and operate three flying

schools in Montana before World War II. "When Pearl Harbor came we went into the Army Air Corps as flight instructors. During that time I married my Glasgow, Montana, sweetheart Elaine Newton in Las Vegas."

Gene's dad died suddenly in 1943, and Gene and Elaine returned home to operate the

"Dad's regular journey in a heavily loaded buckboard to scattered sheep camps would take a week or more. One day, I gave him an aerial tour of all those camps, landing at some. We were back at headquarters by lunchtime."

family ranch. In the late 1940s, as Gene's brothers returned, Gene and Elaine sold them their interest in the Etchart Ranch and together built another substantial ranch, the Hinsdale Livestock Company.

"Dad was innovative," son Joe says. "He

developed one of BLM's early Allotment Management Plans and one of the first silage operations and automated feed yards in the Milk River Valley." His beef promotions were also creative. "The Montana Stockgrowers sent the cellar-bound 1959 Yankees Montana steaks," Gene says. "Steaks they had for breakfast before a sweep of the Baltimore Orioles started their run back into contention."

Family, ranching, flying, history and story-

CLOCKWISE FROM LEFT: Gene and Elaine at Carroll College in 2010 with, from left, John, Jacque, Michele, Janeen and Joe. Gene was a student there in the 1930s and was awarded an honorary doctorate in 1976. ➤ Leonard, Mitch, Ferne, Gene and Mark with Basque parents, John and Catherine, in 1940. ➤ Gene with favorite horse, Croppy. ➤ Gene and hunting buddy O.E. Markle with some of their airplane coyote harvest, winter of 1939-40.

telling are Gene's loves. He maintained his flying status until age 95. He interviewed and recorded old-timers as far back as the 1960s. Those tapes became the basis for several books he has compiled about ranching, flying and Montana lore.

Gene and Elaine still live in Glasgow and celebrated their 72nd wedding anniversary last May. "We have enjoyed some 70,000 meals together, most of them beef," Gene says. "That recipe is a good one for longevity." Gene's younger brothers—92-year-old rancher/pilot Mitchel and 86-year-old physician Leonard—are more proof of that. Gene credits his success to his faith, to Elaine's support, to never asking his crew to do something he couldn't do, and to his father's insistence that he treat others as he'd want to be treated. Oldest daughter, Michele, and her husband, Steve Page, still operate the original Etchart Ranch.

"Dad is a proud man," Joe says. "Proud of his family, his Basque heritage, the country he calls home, and his chosen profession, rancher."—*Joe Etchart (son)*

Harry Hanson, 96

Still fiddling around.

Harry Hanson was born Aug. 24, 1914, in a Norwegian settlement in northern South Dakota. As a small child, he learned to play the fiddle. He tried to make a fiddle from a cigar box and sold Cloverine Salve to earn enough money to buy a violin from the Montgomery Ward catalog for $2.50.

"All the men in the area played and I guess I learned fast 'cause at the age of seven I was taking turns playing at house dances. I was so small that I had to stand on a chair when I played." It is said that he was as good or better than most of the adults.

"My family was real poor

and I grew up in hard times. I had to quit school after the eighth grade and go to work. One fall, a friend and I hitchhiked to Iowa to pick corn. I got a job on a farm near Hull and got paid one-and-a-half cents a bushel. A wagon held 40 bushel and I could handpick two loads, or 80 bushel a day, so I could make $1.20 a day, which was good money."

Those were long days and besides picking and unloading the corn, he had to harness and care for the horses morning, noon and night. "The folks I worked for were real good to me and wanted me to stay on after the corn was harvested but my friend wanted to go home and I decided to go with him."

Harry remembers when his father and others had to sell their starving cattle to the

government because of the drought. "They dug a big trench, shot the cows and buried 'em."

He entered the Civilian Conservation Corps at the age of 17 and first worked in the Black Hills thinning trees. Then, he helped dig a ditch from Lake Poinsett to the Sioux River for drainage. "It was seven miles long and it was all done by hand. We got $30 a

FROM TOP: At 96, Harry entertains at the Cowboy Picnic in Gordon, Neb., in June 2010. ➤ Civilian Conservation Corps camp in the Black Hills in 1934. ➤ Young Harry holds the workhorse behind parents, Blanche and Engval, after a day's work. ➤ Harry working for the CCC. ➤ Harry, center, with parents and grandparents. From left: Norwegians Matia and Hendrik Hanson, Engval and Blanche, and little brother, Albert.

month. They sent $25 home to my folks and I kept $5."

When Harry got out of the CCC camp, he went home and got a job working for his Uncle Martin [Hanson] who had a place along the Missouri River. "I worked for $10 a month and my room and board during the winter months. We cut firewood with a big handsaw. He was on one end and I was on the other. He sold it for two or three dollars a

wagonload."

Harry took jobs wherever he could. He eventually got work as a mechanic in Valentine, Neb., and was soon transferred to Gordon, Neb., as a shop foreman. It was there that he met his life partner, Ruth Kayton, the daughter of an area rancher and dairyman. The two were married in January 1938. "I was making $15 a week and we burned cut-up tires for heat. We were pretty hard up, but my wife knew how to make do."

In the early '40s, Harry purchased his first

"One fall, a friend and I hitchhiked to Iowa to pick corn. I got a job on a farm near Hull and got paid one-and-a-half cents a bushel. A wagon held 40 bushel and I could handpick two loads, or 80 bushel a day, so I could make $1.20 a day, which was good money."

land. "I bought that first quarter for $25 per acre. I planted rye and it got up to $3 a bushel and I made enough off that crop to pay off my land." Harry began putting together his farming operation. "I always credit President Roosevelt and his programs—like the CCC camps, Federal Land Bank, and others—for helping folks get on their feet after the Depression."

Harry continued farming, trucking, running a repair shop, and raising a family, but he also managed to attend fiddle contests. He has won more than 250 championship trophies, including national awards.

After 72 years of marriage, Ruth died in February 2010. He still lives in their home and keeps an active interest in his farming operation. Beef is still Harry's food of choice and he often jokes, "If I would have known I was going to live this long, I'd have taken better care of myself!"—*Yvonne Hollenbeck*

PHOTOS COURTESY HARRY HANSON

Kay Williams, 94

She couldn't drive a car but
she could drive a team.

Catherine "Kay" Piskac was born into an Irish-Bohemian family in Chicago, Ill., in 1916. Sweet and refined, Kay makes her home in Perkins County, S.D., in the breaks of the South Fork of the Grand River. Her simple, tidy house overlooks a deep, rugged draw. "I love my draw," she says tenderly.

Kay's husband, George Williams, a rancher known for his quiet ways, passed away 10 years ago. The two had made their life on the ranch homesteaded by George's mother.

"I was born and raised six blocks east of the Union Stockyards in Chicago, which is how I first learned of my love for the country," Kay says. "You could always tell when it was the day for the cattle—the smell would go all the way to the lake."

Her parents were deaf-mutes and Kay often served as a "voice" for them. "I went with Dad anytime he needed an interpreter." She and her three older siblings learned sign language before they learned to talk.

Kay recalls a time when the family was living with friends in Aurora and her father walked 60 miles to Chicago in search of a job. He found one as a tool-and-die maker and the family returned to Chicago. Groceries and electricity were obtained on credit and sometimes the bills went unpaid for a spell. "It was a good life," she insists. Every payday her father would take the kids to supper and a movie. "Dad enjoyed the movies because back then the words had to be read on screen." Kay's interest in the country grew as she viewed the exaggerated romance and heroism of those Westerns.

As a teen she worked at a candy store and boxed brochures for the 1933 World's Fair. "I started high school several times but we could never keep up with the expenses for bus fare and lunch."

She married before age 20 and had two children, Jack and Karen. The marriage soon ended and after working for a while in California Kay moved in with her older sister in Chicago and got a night job at *Time Life* magazine. "Mary could watch the kids at night and I could earn money correcting subscription lists."

While there, Kay met a ranch girl named

"There was no electricity or running water...but life was no harder on the ranch than it had been in Chicago."

PHOTOS COURTESY KAY WILLIAMS

CLOCKWISE FROM TOP: Kay, 2, Clarence, 5, Johnny, 11, and Mary, 13, in 1918. ➤ Kay, 17, at the World's Fair in Chicago. ➤ Kay today. ➤ Kay and George with, from left: Corinne, Warren, Holly, Jack and Karen.

didn't know what he'd do when I left."

Kay knew what he meant. "I loved him but I didn't want to break his heart if I couldn't take ranch life." She wound up in Rapid City, S.D., alone with her children for several days. "I decided to send George a telegram—'I'll take you up on your offer'—but I didn't realize that he used a middle initial because he was named after his grandfather. Grandpa George didn't figure that telegram was for him," she recounts with a giggle.

George and Kay were married July 11, 1942, less than a month after she had left Chicago. There was no electricity or running water in George's two-bedroom house and she cooked on a woodstove, but life was no harder on the ranch than it had been in Chicago. "I grew up poor. I went through hard times with my family."

Kay and George had two daughters and a son: Corinne, Holly and Warren. Kay didn't know how to drive a car and she and the kids often drove a team to the neighbor's house to catch a ride to church. The children attended several one-room schoolhouses, staying the week with George's father or the teacher and returning home for the weekend. One Friday, George, a man of few words, told Kay, "If you're going to get the kids, you're going to have to learn to drive," and she set off alone to do so.

In the 1980s, Kay obtained her GED and was told that her score was one of the highest the tester had seen. A woman of faith, she is independent and sharp-witted and enjoys visitors. Now her early days in Chicago seem to have been just a short pit stop on the road to a rendezvous with her real-life cowboy.—*Carrie Stadheim*

Eva. In June of 1942, Eva invited Kay and kids to western South Dakota for a two-week vacation. Kay and Eva's brother George became close, walking and talking for hours. He proposed after just a few days. "He said he

We've Got Your Back!

© Cynthia Baldauf

Don't be scared, Bree. Dad, Dan Coon, and cousin, Dennis Kirkpatrick, run alongside to ensure the outcome. Dad can be overheard saying, "We gotcha!" They eventually lifted Bree to safety and she quickly called for a "reride" at the Strowbridge Ranch in Wisdom, Montana.

SPRING 2013

A Sharp Knife & a Shoelace

Mick Marvel was brought into the world by his great-grandfather, Tom. It was no problem because Tom had helped calve heifers for decades. Words & photos by Mary Branscomb.

Not just everyone can claim that he was delivered into this world by his great-grandfather, but by the time Mick Marvel can walk and talk (if the past is an indication of the future), he will have experienced many marvels.

Mick was born on Labor Day, the final Monday of the annual Elko County Fair in northeastern Nevada. The arena in front of the grandstand was full of riders, many of them Mick's relatives who were competing in stock-horse classes between thoroughbred and quarter-horse races.

That morning, Tom and Rosita Marvel were tired from watching children and grand-children and great-grandchildren—all horse-back—at the fairgrounds for two long days. They had visited the Home Arts building to check out the handmade items of rawhide, silver, wood and leather, and to see examples of quilting, photography and canned food. They had visited and sat on hard benches long enough and decided not to go to town this final day.

"Maybe it was someone upstairs who told me to stay home," says Tom. "And maybe He sent me back into the house before I was done cleaning the corral. Anyway, it was unusual for me to be inside, but there I was when Rosita called."

At the IL Ranch, 100 miles away, Rosita's grandson Sam had finished seeing to his cattle and was in cell-phone range on the way back home to the Six Bar Ranch when he again called his pregnant wife, Tori. She explains: "Sam and I knew it was my due date, but I was only uncomfortable when he left home at three a.m. I wasn't having pains. He kept calling to check on me during the morning, and about one o'clock, I told him he better come home. Right away he dialed his grandmother Rosita, who lives only a few miles away."

"Nana," he said, "can you go over to the ranch and check on Tori?"

34

Rosita was already there and had told Tori she was not going to leave although six-year-old Ivy was dressed to go somewhere, anywhere, with her. Ivy, busy now in her own bedroom, would peek out occasionally and query, "Is my brother here yet?"

Rosita knew Sam and Tori had planned to have the baby with a midwife in Idaho three hours away because the young mother wanted nothing to do with the hospital experience. Rosita also knew about the birthing process, having had seven children of her own. Toward the end of a bad pain, Tori asked Rosita, "Does it get any better than this?" And Rosita replied, "Yeah...when the baby comes." Then she asked, "Shall I call 911?"

"No," groaned the girl in the throes of a big one. So Rosita called Tom, her husband of 67 years. "Come over here right away," she said. "This baby is coming now!"

Tom has delivered countless calves and colts on his ranches and he doesn't suffer from indecision. He called 911 and gave them Rosita's cell number while he drove to the Six Bar. When he walked in the door, Rosita had a phone in one hand talking to an EMT en route to the ranch and was cradling Tori with the other. Tom grabbed Tori and had pulled her upright when Rosita said: "Grab the baby. Here it comes!"

So he did. Eight-pound little Mick fell into his hands.

"It wasn't crying and I knew I shouldn't swing it by the heels like I would a calf," Tom says, "so I patted its little butt and handed it off to Tori. Then it began to cry."

Rosita relayed instructions to Tom as she spoke to the EMT who was riding in the ambulance with miles yet to go. "Find a string," she said, "to tie off the umbilical cord." Of course Tom couldn't find a string in the unfamiliar house, so he cut off a shoestring to tie the knot.

"Now find scissors," Rosita instructed.

Couldn't find that either, but he had a pocketknife—a sharp one—so he handed it to Tori who cut the cord.

Meanwhile, Sam was breaking speed limits trying to get home. He roared past a sheriff's car that was following an ambulance on the Lamoille Highway. Naturally, the sheriff's deputy roared right after him and pulled him over. Sam jumped out of his pickup saying, "My wife is having a baby!"

"Where do you think we are going?" replied the deputy. "Follow me."

Mick had already arrived when the ambulance, deputy and Sam wheeled into the yard. Ed Licht, who lives in the bunkhouse, and neighbor Carol Buckner, who had seen Tom drive by her house ("He never drives that fast," she noted. "I knew something was wrong at the ranch."), were also waiting breathlessly as Tori and Mick were loaded on a gurney and driven to Elko. Everyone followed the ambulance except Carol, who stayed to clean up after everyone else was gone.

Before they got to the hospital at about three in the afternoon, news of the birth was spreading throughout the grandstand. After all, it was fair time when almost every person in the livestock industry in Elko, Eureka and Lander counties is in town to watch the races, or help children who have 4-H animals, or who themselves are competing in the arena. And nearly all are friends of the Marvel families.

Mick will forever be known as the boy whose great-grandfather, Tom Marvel, delivered him, just like a calf. ∎

Mary Branscomb met the Marvels in 1965 when husband, Bruce, came to Elko, Nevada, as a young vet. Bruce worked for Tom for many years and Mary began to ride with and learn from horseman Tom. They are dear friends.

"It wasn't crying and I knew I shouldn't swing it by the heels like I would a calf," Tom says, "so I patted its little butt and handed it off to Tori. Then it began to cry."

Ivy and baby brother, Mick. OPPOSITE: Tom, 88, holds Mick and is flanked by grandson Sam, Tori and Rosita. Mick's arrival was a big day.

Mary "Mickey" Thoman, 87
Still ranching and riding.

"I was born in 1929. They say that was a bad winter," begins Mary "Mickey" Thoman, who ranches near Kemmerer, Wyo., 50 miles northwest of the Green River. The family operation below the Fontenelle Reservoir has been in business since 1900, when Mickey's parents, Phil and Mary (Planinsek) Ferentchak, came to this country from Austria. It's now in its fifth generation of ranchers.

Horses play a large role on the W & M Thoman Ranches, LLC. "My dad was a great teacher," Mickey says. "He really liked horses, and we did everything with horses, from haying to herding." She went to high school in Kemmerer, and when it snowed she'd have to get there with the horse and sleigh.

"In summer, we baled hay with horses, handled mowers, rakes, push rakes, all of that. We didn't have a big place. We put up 100 tons."

Of course, they had runaways. "One time driving the rake, bees got into them and the war was on. My dad always advised if anything happened to fall off the back of the rake. I did. The horses ended up with a bush in between them, and the rake was almost on top of them."

When her dad was driving the push rake one of the lines broke and the team started running around in circles. "I was supposed to be driving the stacker team and those horses spooked and ended up in the haystack with me. It was exciting at the time."

After high school, Mickey worked in a drugstore in Kemmerer, where she met William J. Thoman Sr. in 1948. They married

and had seven children (two now deceased) who were and still are a part of everything. "Bill was a sheep rancher so the kids would be with us in our sheep camp in the summer. You have to herd sheep to know sheep. They are smart animals."

> **"I was supposed to be driving the stacker team and those horses spooked and ended up in the haystack with me. It was exciting at the time."**

Mickey ran wild horses with Bill. She was the right-hand hazer on horseback while Barlow Call mustered the herd together with an

airplane. "We gathered more than 3,000 horses through the years. We really enjoyed those horses and used quite a few in our business."

She was involved in 4-H with her children, and has been instrumental in Green River Riders for 56 years. "We would have to drive 70 miles to Rock Springs for a meeting. I feel 4-H was very important in making our children successful."

Mickey and Bill bought a place in 1949 in Green River and were there for 35 years. However, in 1980, the U.S. Fish & Wildlife Service served final eviction notice on the Thoman family after the conclusion of an eight-year court battle on condemnation.

"We owned 25 miles of land along the Green River they wanted," Mickey explains. "We didn't want to move and start over again, but ended up moving 15 miles upriver and started rebuilding a ranch."

Mickey has been at that ranch since 1980. Bill was killed in an auto accident in June 1998 but Mickey has continued ranching with three of her daughters. An inevitable buyout in 2016 forced them to give up their forest allotments in the Bridger-Teton National Forest that they had for 40 years. "The grizzly bears and wolves made it miserable. We are still looking for a place to run our sheep."

In 2014, Mickey was inducted into the Wyoming Agriculture Hall of Fame, sponsored by the *Wyoming Livestock Roundup*. She is 87 and still does all chores on the ranch. "We have a few milk cows, beef cattle, and sheep. It's a full-time job for me." If that's not enough to keep her busy, Mickey still saddles up and rides. "I wouldn't do anything else other than ranch. I love it."—*Rebecca Colnar*

PHOTOS COURTESY THOMAN FAMILY

CLOCKWISE FROM TOP: ➤ *Mickey loads bucks during shipping in the fall.* ➤ *Mickey as "Days of '47 Queen" in Kemmerer, Wyo.* ➤ *Driving Duke at the 2012 Sweetwater County Fair, Rock Springs, Wyo.* ➤ *Bill and Mickey on wedding day in 1948.*

Sydney James, 91

Born to cowboy,
educated to teach.

After many years of education, Sydney James ultimately became a high-ranking professor of agricultural economics, first in New Mexico, then at Iowa State University for most of his career, and finally at Brigham Young University, where he retired. But that's not how he started out.

Syd was born in the spring of 1928 to Ferris and Vera Carter James and raised in remote Park Valley, Utah, just a stone's throw from Idaho and Nevada. "My parents both came from large, multigenerational ranching families, but they insisted that their three sons get more education—and as much as possible—starting out in the one-room schoolhouse in Park Valley."

When the family moved to the Muddy Ranch about 20 miles south of town in 1939, Syd and his older brothers, Carl and Lynn, often had to stay with relatives in Park Valley during the school year. Later, when a school bus started running out their way, it was only during really bad winter weather (which there was plenty of in Park Valley) that they had to stay with family closer to the school.

"When I was about 13," Syd says, "and my brothers had moved on to high school in Brigham City, I got to drive the family car about 11 miles each way from the ranch to meet the school bus each day. I didn't mind that."

Red meat was a staple of the James family diet. "We ate lots of venison that was kept in a root cellar, which served as the household refrigerator. Once, when my mother went to retrieve something from the cellar, a rattlesnake had crawled up the door jamb and was dangling from the door frame. It just about scared her to death. She returned with a rifle, and by the time she was done the snake was no longer a threat, but there were also bullet holes throughout the cellar, including a pan of potatoes that could no longer hold water."

CLOCKWISE FROM TOP LEFT: Syd, center, with two BLM agents at the Muddy Ranch in the 1960s. ➤ The James boys, left to right: Carl, Syd and Lynn in 1930. ➤ Syd and his wife, Carolyn, on the mountain at Tibble Fork in 2017. ➤ Syd on Sadie, 1946. The old dirt root cellar is behind him. ➤ Syd James and cousin Glenna May Goodliffe in 1936. ➤ Syd holds Cheryl at the ranch in 1961; Clair and Susan are on Goldie.

As a change of pace from beef, they also ate a fair amount of sage chicken and cottontail rabbits, which were abundant in the fields. When they were mowing hay with teams of horses, Syd's dad would pack a .22 rifle to shoot the chickens as he encountered them.

"As the youngest son, I typically used the oldest, most well-broke team, while my dad used the greenest. One day after he had worked a young team all morning—and my own team was starting to slow way down and I was having a hard time keeping them going—my dad thought I could probably handle the younger team, and he could get more out of the older team."

They traded for the afternoon and everything was going just fine until his dad got off the mower to shoot a sage chicken, thinking that old team wouldn't make a

"Everything was going just fine until my dad got off the mower to shoot a sage chicken."

bobble, "but they took off and ran away with the mowing machine, scattering it in parts and pieces from the field all the way back to the barn."

By the time Syd was 15, he agreed to stay with one of his aunts and her children in Park Valley over the winter. He was to feed a herd of cows and do their ranch chores while his uncle was working away from home. "I went on the condition that I could bring my horse, Sadie."

Syd would ride Sadie to school every day, and then on the way home ride to one of his uncle's stackyards. "There I would harness a team, pitch on a load of hay and feed a herd of cattle before returning home to milk a cow and do other chores."

At 91, Syd still walks a couple of miles a day and volunteers at a local hospital. "Hard work and red meat are a life-saving combination," he smiles. "Those were the days."

—Todd Macfarlane

Double Duty

© Cynthia Baldauf

Ranch wives often have to utilize a combination of skills at one time. Kelly Kirkpatrick is performing the "calf two-step" while packing her daughter, Maddie, on her hip. It's calf wrangling, aerobics and child care all in one. Young Maddie grew to become a beautiful and talented horsewoman in Fishtrap, Montana.

Walter Teagarden, 82

`Cowboys are tough.`

"I always thought cowboys were kind'a like Superman. They can do most anything. They can mechanic, carpenter, castrate, dock tails and immunize their stock. They can fix anything with a little bailing wire, duct tape and a folding knife."

Walter Teagarden, at age 82, has returned to his true lifelong passion...the life of a cowboy. "I suspect I've always been more comfortable around horses and livestock than two-legged critters."

His approach with livestock, and lamebrained kids for that matter, has changed over the years. "Seventy years ago a knot-head horse and a lamebrained kid were treated about the same." They were generally dealt with swiftly with a "two-by-four across the muzzle or a boot in the butt." Today, Walt treats horses with the patience and tact of the original horse whisperer, but his philosophy about dealing with lamebrained kids hasn't changed much!

Walt was born Jan. 12, 1929, in Boone, Iowa. "My dad passed away when I was young. In fact, we buried him on my 10th birthday." A few months later, he was living in Richmond, Calif., with his mother and new stepfather. Within two years, he was gone. "I saved a few dollars and had no use for my stepdad so I jumped a Greyhound bus bound for Buhl, Idaho." It was 1941.

Shortly after reaching Buhl, he discovered The Sport Shop. It was the local saloon where ranchers could recruit hands. Even at age 12, Walt didn't feel he stood out much as he shared a beer with some of the local cowboys, but he almost got in trouble.

"I did have a run-in with a black cowboy during a poker game in a saloon." Walt says the cowboy thought he was helping a wrangler cheat. "I explained I didn't know the game and was trying to learn." Walt carefully backed out and apologized for his lack of discretion. "I'm sure I dodged a butt whoopin."

Hired his first day in town by a local rancher from the Mellon Valley, Walt was stunned by the bunkhouse. "It was a shack with about four or five rows of bunks fitted with bedsprings and no mattresses or blankets. My first impression was, damn, these

CLOCKWISE FROM TOP LEFT: Thelma, Walter Junior and Walter Senior in Boone, Iowa, 1931. ➤ Walter with registered quarter horse Leena, 2009. ➤ Heartache's Frosty gets washed in Browns Valley, Calif., in 1963. ➤ Walter at 18 months in 1930.

"I did some of my best work while sleeping on horseback or while driving the boss' Caterpillar tractor. To everyone's displeasure, the tractor and I ended up in an irrigation ditch."

cowboys are tough." He didn't sleep much the first night and the next morning the cowboys had quite a laugh about the new kid. "They had forgotten to tell me to get some straw from the barn and pile it on the springs." The next night it was more comfortable.

"On my first trip to town I went to a canvas shop and bought a bedroll." He learned that cowboy bedrolls consisted of a piece of waterproof canvas about twice the size of a normal twin-size sheet. They would lay canvas on the straw and then place a wool blanket on top of the canvas. The cowboy would lie on the first blanket, cover himself with a second blanket, and then roll up in the rest of the canvas. "The bedroll was not the tight little package you see in cowboy movies. They were actually real long and would hang plum near to the ground behind a saddle."

During cattle drives he was assigned the job as the herder and his partner was the camp tender. The tender drove the wagon, cooked, made sure they camped near water, and generally was the boss, whether they were driving 500 sheep or 200 head of cattle. More often than not they drove livestock to and from leased government land in the high desert. The tender led in the wagon while the herder followed along on his horse picking dust out of his teeth. Because of the long hours in the saddle, it was pretty easy to fall asleep in the early morning hours as the sun was starting to come up.

"I did some of my best work while sleeping on horseback or while driving the boss' Caterpillar tractor. To everyone's displeasure, the tractor and I ended up in an irrigation ditch."

In 1948, Walt joined the Marine Corps and was stationed in Southern California where he met Ruby. They got married and raised four kids. In 1963 they moved to Browns Valley in Northern California. Walt worked as a barber and a mechanic but also raised some horses, including an Appaloosa racehorse named Heartache's Frosty. That horse won the California Derby in the late 1960s.

Walt and Ruby celebrated their 50th wedding anniversary in 1998, and it was pretty tough on Walt losing Ruby to cancer a few years ago. But he has rediscovered himself.

"I still feed my cows and prized bull," he says. "And I even talk to my horses."

—*Jack Teagarden*

Barry Naugle, 88

Geezers and wood ticks.

Barry Naugle was born in New York City in 1924. He was a rebellious boy and prone to antagonizing his brothers, so at the age of eight he was sent to a military school where he boarded for four years. He was almost always on the punishment squad. "Why, if I couldn't eat it, kick it, or fight it," he says, "it was no good to me."

Barry changed his ways one summer when he was sent to a guest ranch in Wyoming and he worked with the cowboys and ranch hands. "Once," he says, "I spent a coupla weeks herding sheep with a Paiute Indian. We couldn't speak each other's lingo but we got along good. I liked him."

Barry spent many summers there. "The hands treated me real well. They never talked down to me like Easterners do to a kid."

World War II started and in 1943, at age 18, he joined an infantry outfit. He fought through France, Belgium and into Germany. He was wounded twice. "That was terrible stuff," he says. "Let's skip it."

He wandered to the West at 21, with no job skills. In Wyoming, he went to work at the Z-U Guest Ranch in 1946. That year he met 17-year-old Jackie Chapman, and they married in 1950 after she graduated from college. Barry says, "Well, I was slow courting her, but you educate a girl and you educate a family. I sure was not educated."

They left the Z-U with a car and a cowboy bed. Several ranch jobs and years later they realized they were not getting any closer to buying their own place so they drove to Anchorage, Alaska. They worked all summer and fall for geologists in a dozen camps. Jackie was camp cook and Barry was camp tender. They saved every dime they made.

In 1954, they put a down payment on a 400-acre place in Washington state. "Oh man," Barry says, "did I get fooled. Winter was mild and I thought this was an easy country after Wyoming. Spring came and with it a cold wind. The wind quit in August.

The place was a rock orchard. Not good."

It was eight years of "root, hog, or go hungry." And then four kids sat at their table. "Looked like a bird's nest with four little open mouths waiting to be stuffed."

They sold at a profit and bought another place. They fixed it up and turned it green, sell-

PHOTOS COURTESY NAUGLE FAMILY

CLOCKWISE FROM ABOVE: Jackie and Barry in 2007. ➤ Sid and Amie Reynolds of the Z-U Ranch with Barry Naugle (right) in 1946. The stagecoach used to take visitors through Yellowstone Park. ➤ Barry in 1935, starting his third year at New York Military Academy. ➤ Jackie watches Barry on Pogo in July 1951 outside their first home on Payt Honson's ranch, the GL. ➤ Left to right: brother, Jay; mom, Carolyn; and Barry in 1936 at the Leckie Ranch, a cattle and sheep outfit.

It was eight years of "root, hog, or go hungry." And then four kids sat at their table. "Looked like a bird's nest with four little open mouths waiting to be stuffed."

ing it after several years. "Dang near lost the whole shebang once. Got leptospirosis in the cows and lost a lot of calves. Couldn't pay our bills come fall. But we were trusted and our loan people gave us another chance and we skinned through."

Barry took a correspondence course in math and surveying, giving him a spare trade to fall back on. He says that he had good jobs with the Soil Conservation Service, the state, and at a dam across the Columbia River. "But we always had a ranch of sorts and up to 200 head of steers during the year."

In 1990, they sold out and moved to Idaho where they bought bare ground near North Fork. They spent a summer living in a tent while they built a log cabin. Barry claims this was one of their best times. "We always worked close together and enjoyed it. Work was fun again."

They did all kinds of small construction jobs under the fictitious name Geezer Construction Co. They cut firewood and sold a cord at a time as the Wood Tick Company. Barry says: "One lady insisted that we do fancy rock work on her house's foundation. 'Lady,' I said, 'I don't know diddly about rock work.' But she said, 'You can do it.' We did and it looked okay."

Due to health problems they moved to Emmett, Idaho, in 2007. Barry now keeps busy vegetable gardening, making jigsaw puzzles, and writing stories for the weekly *WesternAg Reporter.*

"You want me to sum up our life?" asks Barry. "Well, most things we accomplished was done by brute strength and awkwardness, but it was good."—*Jackie Naugle*

Good Boys

© Todd Klassy

Rafe Mattison pets his horse and says, "He's a good boy." During a busy time at branding, Rafe's mother, Lu, lets him sit on his horse in the safety of the corral adjacent to the branding action. It won't be long before his legs grow enough to fit those stirrups at the Elliot Ranch in Wisdom, Montana.

James Sheridan, 93
Only a few mishaps.

After graduating from high school, Jim Sheridan wondered what lay beyond the horizon of his tiny hometown of Almo, eight miles north of the Idaho/Utah border where only 140 people live today.

"I went off to Portland's Benson Polytechnical School for six months to study mechanics and got so homesick for this place, I came back and never left," Jim says of the Raft River Valley. This is where his father and grandfather had grazed their Hereford cattle in lush meadows and watered them in streams that flowed from snowcapped peaks.

"This is home," he says, strolling around the Sheridan Ranch, which his son Cordell now runs a few miles west of town. "My granddad homesteaded over there in 1878," Jim says, pointing to a nearby distinctive local landmark: towering turrets of limestone dubbed Castle Rocks. "Now it's Castle Rocks State Park."

He stops at a weathered gray log cabin. "Right here is where I was born on July 26, 1918." Across from the cabin, a board-and-batten barn stands. "We still use it. It used to be the livery stable for the freight wagons that came from Kelton, Utah. It stood in Almo, until my dad took it apart and hauled it here by wagon and put it up. I used to milk 16 cows in here."

In addition to Jim, his parents, Cecil Albert and Eva, had three daughters, "who spoiled me rotten," he says, grinning. Not so rotten that he didn't learn to work hard and deal with the unexpected.

"I had just turned 19 when Dad died of liver problems," Jim says. "We took him to Burley to the hospital, and he never came home. My sisters had married and moved away, so I was left with this place to run. The neighbors helped us out."

Jim tended to his Herefords, married Dorothy, a teacher, in 1948 and helped raise their children: Marcia, Sherry, Cordell and Marlon.

PHOTOS COURTESY JAMES SHERIDAN

CLOCKWISE FROM LEFT: Jim rides a "bronc," a roadside attraction, while en route to Jackson, Wyo., on a family vacation. ➤ From left: Jim Sheridan, Arlo Kyle, Lucretia Kyle, Cecil Albert Sheridan and his wife, Eva. ➤ Jim's dad dismantled this livery stable, moved it to the ranch, and reassembled it. It's still in great shape. ➤ Jim's parents, Cecil Albert and Eva, relax on the ranch in the early days. ➤ Jim today.

> **"There was a grain chopper parked here, and I reached in to flip out a wire. It ate my hand pretty good. I've had surgery on my knee and for cataracts. Then in 2007, I had a heart attack and three bypasses, but I can still live a normal life and eat some beef."**

"In the summer of 1971 we needed a little extra money and I had a chance to work on a survey crew with the Bureau of Land Management," says Jim, who juggled running the ranch with holding down summer jobs for the next 36 field seasons with the BLM and U.S. Forest Service.

Five years later, when he was 58, he started a new summer job: working as a fire prevention patrolman for the Sawtooth National Forest. In 1985, at 67 and when most people consider retiring, the Forest Service hired him to work in the fire lookout tower at nearby 9,265-foot-high Mount Harrison.

"Any place I saw from up there, I had been to on horseback, foot or vehicle." When he finally retired in 2007 at age 89, he was the oldest fire lookout nationwide.

Eventually Cordell began running Sheridan Ranch. "He decided to switch to Gelbvieh and Angus, and he's done a fine job of it," says Jim of his son's focus on improving herd genetics and selling a crop of bulls yearly.

Jim admits to a few mishaps during the past decades, plus a few pesky physical ailments, but they are a source of entertaining stories rather than self-pity. His right index finger is missing. "There was a grain chopper parked here, and I reached in to flip out a wire. It ate my hand pretty good. I've had surgery on my knee and for cataracts. Then in 2007, I had a heart attack and three bypasses, but I can still live a normal life and eat some beef."

He uses a ski pole when he walks around to help him maintain his balance. "It's standard operating equipment for me these days. I may not get around as good as I did 40 years ago, but I still get around pretty good for 93."—*Diana Troyer*

Learning Them Early

© Skye Clark

As Jake Wolaver holds the heels, Gabe Clark, cowboss of the Grindstone Ranch in Daniel, Wyoming, brands a calf while holding his son, Jak.

Betting on Yourself

A firm grip and hard-earned calluses.
By Patrick Dorinson

When you shake the hand of anyone who works the land or cares for livestock, you can tell a lot about that person. During this brief exchange, the nicks, scars and bent fingers each tell a story of a life of hard work. Ben and Stella Elgorriaga of Madera, Calif., have that strength of character which comes from having been tested every day by everything man and Mother Nature could hurl at them.

The Elgorriaga family and their ranching operation are like countless families across America, who quietly go about their chosen work, raising their children, struggling at times to make ends meet, never looking back, always with an eye to the future. As Ben says, "Ranching is betting on yourself."

Esteban Bibiano Elgorriaga left the town of Enderlaza in the

Tested every day by everything man and Mother Nature could hurl at them.

Basque Pyrenees in Spain to join his brother in California to work as a sheepherder. On May 4, 1920, he disembarked from the steamship Leopoldina and set foot from Ellis Island to begin the long journey west.

Not all Basques had a sheepherding background, but they shared a strong work ethic, rural roots and a willingness to endure the hardships and isolation associated with that life. Once settled, the immigrants wrote home about their lives in America, which offered more opportunities than the Old Country, perennially mired in poverty. They would work hard, save money and eventually return home to buy their own land or start

a business. Most stayed to follow their dreams in the United States. So it was with Esteban.

After he had worked with his brother for a while, Esteban started building his own flock until the Great Depression wreaked havoc on the world economy. Rural America was particularly hard hit as the Dust Bowl in the Midwest turned farmers into beggars

Great Pyrenees guardian dogs help sheepherders protect flocks from predators out on the open ranges. This was a very good, wet year in 2008 on Ben and Stella's home ranch in the Ciervo Hills in California's coastal range.
LEFT: Esteban Bibiano Elgorriaga Sr. with wife, Alice, and their two children, Esteban Bibiano "Ben" Elgorriaga Jr. and Benita in stroller, 1952.

and migrants. "My dad was able to hold onto his sheep as he staved off foreclosure," Ben says. "He survived, and times were about to get better."

During World War II, the military needed vast quantities of wool to outfit the large number of soldiers, sailors, airmen and Marines who were headed overseas. Congress adopted a wool incentive payment to encourage sheep production. That generated about $20 per ewe and was good news for sheep ranchers. (The wool incentive was discontinued in 1995.)

The postwar years were prosperous and happy for Esteban. In 1949, at the age of 55, he married Alice Mickelsen, who was much younger than he. They had two children, Ben and Benita. "My father wanted to be established financially," Ben says, "before he took on the added responsibility of a family."

In the late 1950s the family prospered, and Esteban served as vice president of the California Woolgrowers Association. He was one of the founders of the California Range Association, which later became the Western Range Association. And in 1961, 40 years after he had first set foot in America, Esteban Elgorriaga retired from the sheep business. He had gone from being a penniless Basque immigrant with a dream to one of the most respected sheepmen in the country.

Ben had always worked with his father, but he was studying in college and didn't get into the business right away. His father died in December 1970 at age 75. Soon after, Ben met Stella Guyette, who attended the same Catholic high school as Ben's sister, Benita. Stella was from a family of 12 children and, like Ben, she was used to hard work. Their courtship consisted mostly of writing letters.

In May 1972, Ben and Stella got married and decided they wanted to get back into the sheep business. Ben worked for the Talbott Sheep Company for four years, then in 1977 the couple bought their first bands from Bill Chandler of Dixon, Calif. "We probably couldn't have picked a worse time to strike out on our own," Ben says. Inflation was high and interest rates soared to around 20 percent. But they still took the risk, betting on themselves.

"By the early 1980s we were barely hanging on," Ben says. "At one point we were making our interest payments in loads of lambs." Tough times got tougher and they tried a joint venture with old friends. It too succumbed to the times, but in 1989 Ben and Stella, along with their children, started Elgorriaga Livestock. Today the outfit consists of 3,500 ewes and 400 head of cattle; they also custom harvest 5,000 acres of pistachios each September.

Elgorriaga Livestock strongly supports the H-2A visa program that allows ranchers to bring sheepherders from Peru, Mexico, Chile and Bolivia for three-year periods to work tending sheep throughout the West. They are still members of the Western Range Association, which handles the arrangements to bring these necessary workers to America.

"Without the H-2A visas and the workers they bring," Ben says, "many sheepmen would be out of business. If you are looking for an American to do the lonely work of a sheepherder without all the creature comforts of a modern society, you will be looking for a very long time."

Unlike sheep operations in other parts of the West where the bands graze on public lands isolated from population centers, the Elgorriagas and others in California's Central Valley graze sheep in harvested fields of alfalfa and wheat stubble, melons or broccoli, gleaning them of the residue left over from harvest. Crop-residue grazing—or as Ben says, "following the feed from farm to farm and ranch to ranch"—requires regular movement of the bands.

The Elgorriaga story comes down to one word—family. Ben and Stella have raised five children, all of whom work for the family business. "We have been given a gift and we want to hand it down to our children and grandchildren," Ben says. "We need to create a place for them. But if they want to come work [on the outfit], they have to fit.

"This land is a source of freedom for the Elgorriaga family. It allows us to work and raise our family as we want to. Two or three generations from now our grandchildren and great-grandchildren might not want to keep doing this work. It is hard and demanding. But the ranch will still be common ground where we can come together as a family." ∎

Patrick Dorinson is a writer and radio commentator from Folsom, California.

Esteban Bibiano Elgorriaga as a child (sitting on right) with his parents and siblings in Spain, ca. 1905. BELOW: Elgorriaga family at the 150th anniversary of the California Woolgrowers Convention in July 2010. Ben and Stella (far left) pose with their five children and 10 grandchildren. Benita (far right) gave birth to their 11th grandchild shortly after this photo was taken.

"Okay, James, I love you, too, but keep in mind I have work to do."

James Campbell offers another cuddle before his border collie/kelpie, Zip, is called to work at the family ranch in Lonerock, Oregon. © Deanna Campbell

Gerald Turner, 88

Learning from the best.

Pictures of John Wayne decorate the walls of Gerald and Ella Mae Turner's modest ranch house just west of Pawnee, Okla. That should be a clue to how the veteran rancher views life. Another clue might be the old barrel propping up their mailbox at the end of the driveway with ominous letters on its side.

"We were having trouble with people destroying my mailbox," Gerald says. "I put a barrel under the mailbox and painted TNT on the side so they would think it would explode if they hit it. Haven't had any troubles since."

Like John Wayne, the 88-year-old doesn't take much guff from city-slicker salesmen or cowboy wannabes. As a young boy growing up on the Oklahoma prairie in the late 1930s, he learned how to rope and ride from wild-west-show legends Pawnee Bill and Mexican Joe, and famous rodeo cowboy, Ben Johnson.

"Pawnee Bill always welcomed me to his ranch," Gerald says, "so I hung around there whenever I could. Mexican Joe worked for Pawnee Bill as ranch foreman and Ol' Joe was the best roper around. He could rope cattle with his foot! He even showed me how to make a rope. Ben Johnson was the best steer roper I ever saw."

When World War II erupted, Gerald enlisted in the U.S. Army Air Corps and worked as an airplane mechanic servicing B-17 bombers. While stationed in Belgium, he and his buddies—all tired of eating Army chow—sneaked into a nearby town to dine on steak.

"Best steak I had eaten in a long time," Gerald says. Even after he discovered it was horse meat, not beef, he didn't complain. "I was plenty hungry."

During that time, he also ate his share of carrots and beets. "Beets are no substitute for beef," he says. "Maybe that should be on a

> ## "Pawnee Bill always welcomed me to his ranch, so I hung around there whenever I could. Mexican Joe worked for Pawnee Bill as ranch foreman and Ol' Joe was the best roper around. He could rope cattle with his foot! He even showed me how to make a rope. Ben Johnson was the best steer roper I ever saw."

bumper sticker! If I was a vegetarian, man, I'd really suffer. I like fruits and vegetables, but without meat, I'm still hungry."

Gerald's roots are planted deep in this land once known as Indian Territory. "My grandfather moved to Oklahoma to teach Pawnee Indians how to farm." Wagon tracks are still visible in the hay meadow north of his house. He estimates that the trail dates to the 1860s, and was perhaps once part of the Chisholm Trail. This particular portion goes to White Eagle in neighboring Kay County, where his great-grandmother was postmistress.

Today Gerald continues to operate a beef cowherd and still frets over first-calf heifers during snowstorms, just like he has for most of his nearly nine decades on earth. He gets plenty of help from his wife, three sons, and two daughters. Gerald and Ella Mae have been married 59 years. "The secret to marriage," he says, "is to take 40 percent and give 60 percent."

In 2012, Gerald received Oklahoma Farm Bureau's Legacy Award, presented to those who have served as an inspiration and mentor to young agriculturalists.

"When I was a

CLOCKWISE FROM TOP: Gerald tells youngsters, "Never give up, no matter how hard things get." ➤ Gerald served in Belgium during World War II, working as an airplane mechanic servicing B-17 bombers. ➤ Posing with his young cousin in 1942 in Pawnee, Okla.

young man, I learned from some of the best cowboys in the country, and I hope I have passed on some of that knowledge along the way," says the weathered cowboy, who isn't necessarily talking about roping and riding skills. "I tell these young guys to never give up, no matter how hard things get. Hard work will make you a success."—*Sam Knipp*

Larry & Hazel Clay, 86 & 84

Circle C cowboying
and cooking.

"The spring of '44 we made one drive from Pittsburg to New Meadows," Larry Clay says. "Cattle were strung out for five miles going up the highway with six to eight riders." He cowboyed for the Circle C Ranch for 42 years and easily remembers the time before U.S. Highway 95 became busy with traffic and the ranch used this stretch from New Meadows to Riggins as a cow trail. They moved the cattle

up and down that trail, twice a year, "whether you wanted to or not."

Larry and his wife Hazel laugh at antics that happened while trailing cattle. There was the drive when a bull saw its reflection in a fancy car and charged it. There was a time a monkey scattered the herd. "On another drive," Larry laughs, "an old gal had goats in Riggins. Those goats scattered cattle through freshly hung laundry. A girl gave me a heck of a cussing for tearing down her clothesline."

Hazel served many meals to cowboys, ranch hands, haying crews and visitors in the old days. "It was breakfast, dinner, and supper back then." When the cowboys were on the trail, she fed them hot meals on wheels. "I'd take the meal in the car and catch up with the herd," she says. "A cowboy would give his horse to another man to lead, then climb into the car. He'd get the chance to warm up and eat while I followed the herd. He'd switch with the next cowboy and so on until everyone had been fed."

Larry was born in Meadows Valley on

FROM TOP: Circle C home ranch, New Meadows, Idaho, June 17, 1963. Cowboys, from lef: Larry, Jerry and Frank Clay and John Stover. ➤ Larry hung up on saddle bronc at Grangeville Border Days Rodeo, July 1951. Pickup men are George Gill, left, and Frank Clay. ➤ The Gill family on Christmas Day in 1942 at Rapid River. Left to right: Charles, Charlie (father), Hank, Goldie (mother), Bill, Hazel and Marvin. ➤ Larry and Hazel pose in front of their retirement house in New Meadows, Idaho, Aug. 11, 2011.

Sept. 1, 1925, the fifth of six sons born to Henry and Katie Clay, joining one sister. Upon retirement, Larry and Hazel bought a house in New Meadows, an area known for harsh winters. "To show how smart I am, I've lived in New Meadows all my life, then when I retired I bought a house where the snow slides off the roof and piles up on the front porch!"

Hazel was born to Charles and Goldie

(Troeh) Gill on April 27, 1927, in Grangeville, Idaho, the only daughter with four older brothers to watch out for her. She still owns land her grandfather homesteaded at Gill Point in 1885.

"My first job was packing a bucket of water a day about a hundred yards uphill to the schoolhouse for 50 cents a month," she says. "I wanted a watch so bad I saved and saved until I got one. That watch was like an old alarm clock hanging on my wrist, but I was proud of it!"

Hazel attended business college in Walla Walla, Wash., then went to work in a Grangeville bank. "I met Larry on the street in Grangeville. Then I went to a Riggins Rodeo meeting with some friends and met him again." They married on Aug. 24, 1952, and Larry says, "She's the luckiest gal there is to get me!" Their son Rick was born June 17, 1955.

> "My first job was packing a bucket of water a day about a hundred yards uphill to the schoolhouse for 50 cents a month. I wanted a watch so bad I saved and saved until I got one."

Larry helped form the Salmon River Cowboys Association to carry on the Riggins Rodeo in 1948. Hazel was the rodeo treasurer for many years. In 1949, Larry won All-Around Cowboy. In 1998, they served together as grand marshals and are still honored as charter members. In all his years of rodeoing and cowboying, Larry says, "I never was busted up real bad. Landed on my head a few times. I was 77 before I had a broken bone."

He remembers some wrecks well: "Above Pollock, the river was faster than I thought it was and flipped my horse out from underneath me. I was able to get a hold of his tail and come out." Larry knows he was lucky but his brother Ed wasn't. Ed drowned in the Snake River crossing cattle during high water. "We searched 22 days before finding his body."

The Clays still attend rodeos and enjoy socializing. Just last year, Larry turned his old mare out to pasture on Hazel's childhood farm. She says, "I never used to pay attention to wheat prices, but now that my interest in the old family place supplements our retirement, I keep a close eye on the market." —*Shelley Neal*

Evan & Tillie Zimmerman, 87 & 86

Moving east to the Wild West.

For 67 years, since they were first married, making bold, expansive moves has been a way of life for Evan and Tillie Zimmerman. They met while attending Petaluma High School in Northern California. Both had been raised on farms—Evan on a dairy in Tomales and Tillie on a chicken farm in Petaluma. Soon after they married in 1944, they bought a farm and raised milk cows, laying hens and three young sons. The future, chapter two of their biographies, seemed already written, just waiting to be lived.

One May evening in 1950, Evan read a small classified in the *Western Livestock Journal* advertising a cattle ranch for sale. Something in him stirred. On Memorial Day weekend, he and Tillie drove to the Disaster Peak Ranch. The following week they made a deal to purchase it and on June 25 they shocked many by moving to the high-desert ranch on the border between Oregon and Nevada.

"It all happened so fast," says Tillie. "In 25 days we had sold our ranch and assets, had our teeth checked and made sure the boys had all their shots."

Despite the fact that they moved east, the Zimmermans had landed in the Wild West. Evan had traded his milk cows for beef cattle, work boots for cowboy boots, milking stool for a saddle, and the known for the unknown. He had a lot to learn and he did so by listening, watching and doing. "The first time I joined some neighbors to gather and brand," he says, "a neighbor got me off to the side to let me know that I had my spurs on backwards."

After a few years the operation expanded in two ways. First, by the purchase of the Turner Place just a few miles up the creek, and second, they had two more sons. Tillie says, "I had always wanted to live on a ranch and raise sons."

School for the boys was a constant challenge. The 30-mile dirt road to McDermitt was twisty, rough, strewn with bog holes in the spring and often snowpacked in the winter. A daily trip to town was not feasible and over the years they tried different solutions such as living in town and homeschooling. "One year Ross and Ted lived with my dad and mom in Petaluma," Tillie says. "That was horrible. I missed them too much."

The Zimmerman boys were often the only help Evan needed on the ranch. All five sons wanted to be ranchers, but the Disaster Peak Ranch was not large enough to support five more families.

In late May 1972, Evan and Tillie toured

> ## "The cattle were browse cattle instead of grass cattle. They were almost like wild animals that could fend for themselves. They did not look to us for anything."

PHOTOS COURTESY ZIMMERMAN FAMILY

the vast RO Ranch in central Nevada. By July 1 they had formed the Zimmerman Ranching Corporation with their five sons, put up the Disaster Peak Ranch as collateral and moved in as the new owners of the RO. They ran a couple of thousand head of cattle on about two million acres of BLM and Forest Service land.

Once again, Evan and Tillie faced a giant learning curve. Ranching at the RO was nothing like that at the Disaster Peak. "The cattle were browse cattle instead of grass cattle. They were almost like wild animals that could fend for themselves. They did not look to us for anything."

Evan used a rest/rotation system, which constantly improved the range, and in 1980 was recognized by the Society for Range Management as Rancher of the Year.

In 1978, they bought two adjacent ranches, the Triple T and the Monitor, and with this expansion they managed several thousand head of cattle spread over 3.5 million acres.

"We had a cow/calf operation in central Nevada," says Evan, "and we sent our weaned steers to the Peak where they really stacked on the weight." In the years that followed, high interest rates and environmental calls for removing cattle from public lands induced the Zimmermans to sell the ranches in central Nevada. Each of their five sons pursued new careers, and Tillie and Evan returned to the Disaster Peak Ranch where

CLOCKWISE FROM LEFT: Evan and Tillie with the boys. From left: Arnie, Ted, Ross, Buzz and Dennis. They all wanted to be ranchers but the Oregon outfit wasn't big enough so they moved to the giant RO Ranch in central Nevada.
➤ *The home ranch at Disaster Peak in Oregon.*
➤ *Evan and Tillie's wedding day in 1944.*
➤ *Evan, Tillie and the boys horseback in Oregon.*

they live today.

In 2008, the Nevada Cattlemen's Association recognized Evan for having ridden horseback over 100,000 miles. Today, at 87, he has swapped a horse for a Polaris six-wheeler he calls The Cadillac. "That is some machine," he says. "It goes anywhere a horse could go, but it is more comfortable and heated."

Tillie has also embraced 21st century amenities. At 86, she is a champion at email, and keeps up with her numerous family and friends on Facebook. Plus, she has a robotic lawn mower named Oscar. Watching it operate is an entertaining pastime and "he" fascinates visitors.—*Julie Zimmerman*

Amanda Mina Elder Banister Barrie, 94

She needed refining.

Amanda Mina Elder, born Feb. 24, 1916, was a cowgirl who grew up horseback. "Before I went to school, I used to go horseback," she remembers, recalling rides to help feed cattle or travel back and forth to a ranch owned by her father's cousin, Enos McDonald. It was near Paisley, then and now a small, southern Oregon ranching community.

But it's an old black-and-white postcard that summarizes life on the ranch. The photo shows a young Amanda Mina Elder standing between her grandfather, James Newton Taylor, and family friend Earnest Mathes. She is wearing britches and an ear-to-ear smile, and barely stands waist high to the two men. Alongside are two gutted bucks, hanging upside down. The men are holding shotguns. Amanda also holds a gun.

Growing up on the family ranch outside Paisley in the early 1920s wasn't the path to becoming a demure young lady. At age 94, Amanda looks back on those early years with a soft smile. "I've lived a satisfactory life," she says, relaxing at the home near Lakeview that she shares with her daughter and son-in-law, Diana and Chuck Adams.

The Elder family had moved from Tennessee, where they grew cotton and tobacco, to claim a homestead in Chewaucan Valley in the 1850s. Her parents, Evalyn and John Bringle Elder, divorced when she was only four and she was taken to Ashland—a three-day trip that included overnights in Bly and Klamath Falls—to live with relatives and start school. "There were no paved roads."

She returned to Paisley to start the fourth grade. "The townspeople kind of kept track of all us kids."

Because of her rural lifestyle and her father's involvement with the ranch ("He was looking out for the cowboys. He was gone a lot."), some townspeople were concerned.

"Some of the ladies," she tells, "thought I needed a little culture."

Her sophomore year was at Klamath Falls and Sacred Heart Academy, a Catholic school where she took piano lessons, learned to play

Amanda's dad thought she was becoming too much like a Catholic, so after six months she was sent to Southern California to live with her mother. "I had a ball," she says. "I loved it."

CLOCKWISE FROM TOP: A young Amanda Elder, shouldering a rifle, stands between her grandfather, James Newton Taylor, left, and Earnest Mathes in an undated photo. ▶ At 94, Amanda enjoys life in Lakeview, Ore. ▶ Evalyn Elder with daughter Amanda in 1942. ▶ Joe and Amanda Banister in Los Angeles in 1934.

the saxophone and had instruction in china making and oil painting. Saturday nights she and a girlfriend attended dances in nearby Malin. "The big bands came through there."

At Sacred Heart there were regular prayers and her education was church focused. Her dad thought she was becoming too much like a Catholic, so after six months she was sent to Southern California to live with her mother.

"I had a ball," Amanda says. "I loved it."

After graduating from high school in 1933, she returned to Paisley and met Joe Banister, a local barber. They soon eloped. "My dad was very angry. I was supposed to go to teachers' school in Ashland."

The next several years were spent shuttling between Paisley and Southern

California. Joe worked in Hollywood. "The stars of the radio shows would come in the shop and give him tickets to the shows as tips," Amanda says of how they attended a "Fibber McGee and Molly" broadcast. "On Saturdays we went dancing to all the big bands."

The wanderings continued—Alturas, Pasadena, Los Angeles, Napa, Canby and four years in San Francisco—always punctuated by returns to Paisley. During their travels, Diana was born in 1941.

After 17 years of a suitcase marriage, Joe bought a Lakeview barbershop in 1948 and a home in 1950, and life finally settled. Amanda worked 12 years for the California-Oregon Power Company and 20 years for the Fremont National Forest.

Joe died in 1974 and three years later Amanda married Gordon Barrie. They lived in Klamath Falls and traveled until his death in 1984. These days she plays bridge and takes care of her garden. "I like to get rid of the weeds."

The gun-toting little girl shown in the postcard may have grown up, but she has never really been tamed.—*Lee Juillerat*

Stories to Tell

Cowboy Bliss
© Allison Randall

Brooks Randall shares a private joke with his curious horse, Cole, and Charlie the dogie calf that Brooks bottle-fed from two days old.

Start 'Em Young
© Andrea Rieger

Emory Reiger, 18 months, keeps an eye on the folks working the ground at a branding in Ismay, Montana.

Alma "Dickie" Ross & Alva "Happy" Carlson, 90

Old days, old ways.

Most Americans enjoy modern life but there are two old ladies who would be happy to trade most modernity to go back to a rough ranch childhood of nearly a century ago. Alma Ross, known as Dickie, and Alva Carlson, known as Happy, are 90-year-old twins

Dickie and Happy are usually referred to as "the twins" because they dress exactly alike every single day. They were born in 1922 on the family homestead outside of Big Sandy, Mont., the last of seven children born to German parents. "We spoke only German until we started school," laughs Happy. "We learned English quickly in school—but we still sneak in a little German when we're telling secrets."

By the time they were 10, their older siblings had left the ranch and America was in

Our house was a long log room, with a bedroom at each end. We lived together in the middle—without electricity or indoor plumbing. We often couldn't see out because our windows were covered with thick ice."

Life was hard, but they recall those times with smiles. "Our parents made sure we had time for fun, too," Happy says. "I think that was the main thing that made us smile our way through life! We had to work harder than most of today's men, but we were surrounded by the love and care of a wonderful father and mother."

They were five miles from the one-room school. Every morning they'd ride bareback to get there. "When a snowstorm hit, Dad would hitch up the sled, wrap us up in great old beaver coats and take us to and from school. We never missed a day." The girls always wore short haircuts. "Mom didn't have the time or the talent for fancy feminine cuts."

When they got older, their father would hitch up the sled and take them to every school dance. "We both loved dancing," says Happy. "When we couldn't find a boy, we danced with each other. Dancing beats the heck out of video games."

Dickie says: "Youngsters back then invented their own fun. We played the standard games, and created a bunch of our own." They claim no special diets for their long, healthy lives. "We ate good, homegrown food mostly," says Dickie. Did they eat cakes, pies and candy? "You bet," laughs Happy, "and whenever we could get them."

As adults, each married and moved away. But they came back together again when they were widowed. They live in a small house near children, grandchildren, and great-grandchildren, who drive them wherever they want to go. They have a television, but don't like the content so they don't watch much.

"We love to go for walks. We always check out the garage sales. We don't buy much, but it's nice to visit."

When asked the secret to a long happy life, Happy laughs. "Just keep breathing and moving," she says. Dickie adds, "And dancin'."—*Bill Kiley*

© BILL KILEY

"I'd love to have kids transported back about a hundred years for the first 10 years of their lives. They'd be so much happier today if they had lived as we did, in an old log home in the middle of nowhere, caring for animals, working at hard chores."

PHOTOS COURTESY THE TWINS

CLOCKWISE FROM TOP: Adult twins, Dickie, left, and Happy at home in Livingston. They would like to go back to their "rough" childhood and worry about today's youngsters. "I shudder when I see all those great little kids just sitting and staring at the TV for hours on end, or wasting time on video games," Dickie says. "You can tell they're usually buried under a cloud of boredom." ➤ Toddler twins. ➤ Baby twins. ➤ Teen twins.

who live in Livingston, Mont., where they worry about today's youngsters.

"I shudder when I see all those great little kids just sitting and staring at the TV for hours on end, or wasting time on video games. You can tell they're usually buried under a cloud of boredom," says Dickie. "I'd love to have them all transported back about a hundred years ago for the first 10 years of their lives. They'd be so much happier today if they had lived as we did, in an old log home in the middle of nowhere, caring for animals, working at hard chores."

Daughters of Fritz and Auguste Jappe,

the Great Depression. Their father needed all the help he could get, so the girls herded the cows to the river so they could drink. "While the cows slurped," Dickie says, "we fished or gathered berries for dinner." They took over the chores of milking the cows, feeding the chickens, pigs and sheep—and gardening. They grew almost every bite they ate. "Remember, we had to go 15 miles for a bottle of milk or some sugar."

Much of summer was spent getting ready for winter. "Dad would chop down trees and set us to work with a saw and ax, chopping wood for our only source of heat.

Jack Liskey, 90

Change is the only constant.

Jack Liskey is living history. For 90 years his personal story has reflected ongoing changes in the Klamath Basin of southern Oregon and far Northern California. With a life marked by hard work and a willingness to test new ideas, it is also the story of an extended family and its role in transforming undeveloped lands for agricultural purposes.

Jack's history began March 2, 1921, in a tent near Malin, Ore. His parents, John and Philomine "Phil" Liskey, had arrived early in Klamath Basin. John was born in Nebraska, where his parents lived after moving to the United States from Germany and before moving to the Swan Lake Valley near Klamath Falls.

Philomine moved from central California to the Klamath Basin in 1915, when she and a woman friend homesteaded on adjacent parcels in the Swan Lake area. She and John married in 1917.

The Liskeys moved to Malin and, later, the Tulelake Basin just south of the Oregon-California state line. Jack's father raised sheep and cattle. He also farmed, growing grass, rye hay, barley and oats. Life was rugged. They sometimes lived in a tent with a wood floor or a small cabin built atop a wagon.

Philomine supplemented the family's income raising "thousands of turkeys," Jack remembers. Turkey Hill near Malin got its name from the massive numbers the Liskeys and others ran in that area.

"We'd run 'em out in the fields," Jack says. "Once you'd get 'em going, they'd come in on their own."

The Liskeys, including Jack's uncles and their families, also ran, gathered and sold horses and mules. His parents moved to a mostly unsettled area along present-day Lower Lake Road in 1931. Before moving, Jack and his father developed area leases where they planted grain.

Life was rugged. Jim and Philomine sometimes lived in a tent with a wood floor or a small cabin built atop a wagon.

PHOTOS COURTESY JACK LISKEY

CLOCKWISE FROM TOP: Turkeys at Turkey Hill near Malin, Ore., raised by Jack's mother, Philomine. ➤ Jack enjoys old photos in a family album. ➤ Jack "cowboying" on a milk cow. ➤ Jack and Vergie's wedding day in 1943.

"We came out here and farmed, Dad and me. I was about nine. I drove the tractor and Dad sat on the wagon and broadcast the rye. That was our first crop we put in."

Jack went to schools in and near Tulelake before graduating from Merrill High School in 1939 and moving on to Oregon State University, where he met Virginia "Vergie" Hooper. She had grown up in Tulelake. They married in 1943 and he left college and was in officer's training school when World War II ended.

Jack and Vergie have four children: Vickie, Rocky, Bernie and Tracy. Bernie was born on Jack's birthday. "He was Dad's 30th birthday present," Vickie says.

Jack and his father farmed and ranched together, then Jack worked with his sons until he retired.

While drilling for water in the 1950s, the Liskeys struck 178-degree geothermal water. Unsure of what to do with the resource, they experimented. They put frozen cull potatoes into hot-water vats and cooked the spuds before feeding them to cows. Jack later installed greenhouses that Vickie used for raising nursery plants. The geothermal site has expanded with greenhouses used to raise organic vegetables and tropical fish. Future plans include a possible geothermal energy generation plant.

"Dad was very innovative," Vickie says, "and he wanted to see what he could do with the resource."

Rocky and Tracy continue to ranch. Of their 1,500 acres, 400 are used for growing hay, 500 acres serve as pasture for 300 mother cows, and other lands are leased out.

Vergie died in 2008, but Jack continues to live on the ranch. He grows a garden and keeps an eye on the family business.

"Dad still helps us feed calves," Rocky says.

"He was raised here. He's comfortable here," Vickie explains. "He's home."

And Jack agrees, saying, "This is what I know."—*Lee Juillerat*

Help!

Grandma, Put Down Your Camera

© Robin Dell'Orto

Waylon Dell'Orto (six years old) was riding a horse named Tres near Jackson, California, while family and friends were gathering cattle. Waylon was sent by himself to check for cows in a corner of the ranch. When I rode my horse up an incline he was CALMLY holding his horse in the middle of the field. "My saddle slipped, and we need to hurry up," he said. He had me place his saddle back on top his horse and we rushed to catch up with the herd.

Soaked to the Skin

© Rick Raef

Ella Villagrana runs toward the gate to get dry clothes for little sister, Elena, after she tumbled into the creek at the Duarte ranch branding near Beatty, Oregon. Six-year-old Ty Duarte runs alongside.

Evelyn Cuneo, 94

Only got on a horse once.

"That big can of ore samples came out here on the floorboard of the car," 94-year-old Evelyn Cuneo says. "It was too heavy to mail. Came from the Oliver Mine in Michigan where Dad worked." Her parents, Elizabeth and Alfred Penrose, came from England. Her dad found work in mining as did generations of Cornish people since the California Gold Rush.

When Evelyn was a teen, the family drove to Jackson, Calif. Evelyn didn't mind. "I used to walk to school near Lake Michigan during minus 40-degree weather," all bundled up "with just my eyes showing."

Her parents started Penrose Dairy. Little sister Margaret washed bottles. Evelyn and sister Joan made deliveries twice a day in the truck. "I was always late for school. Saw a lot in the early morning and evening around town!"

Those were the days of Prohibition. "Dad would say, 'You hear nothing, see nothing, and don't come home and tell your mother!'" The girls developed the habit of whistling when they approached houses so those within knew it was only the milkmaids.

One time the state milk inspector stopped Evelyn, which was a normal occurrence. He took a sealed pint of milk for testing. Later in the mail, her dad received a blue ribbon from the California State Fair. They still have that ribbon.

Once the kids graduated from Jackson High School, George Sausman, the local Chevy dealer, let the kids borrow his used cars. "A group went to Yosemite. The cars weren't very good but they made it."

When Evelyn and Vernon Cuneo got married in Carson City, Nev., George let the couple borrow a new convertible. "We kept it as long as we wanted. George was very good

Those were the days of Prohibition. "Dad would say, 'You hear nothing, see nothing, and don't come home and tell your mother!'"

CLOCKWISE FROM TOP: After Vernon and Evelyn's honeymoon, friends threw a shivaree, where they banged pots and pans, clanged cowbells and danced. ➤ Evelyn with her children at their annual chestnut roast in Jackson. Back row, from left: Jimmie Cuneo, Loree Joses, Evelyn and Verna Payne. Front, from left: Robin Dell'Orto and Kathleen Brownlie.
➤ Evelyn's grandson JW Dell'Orto goes after stray cow and calf at Jimmie Cuneo's calf marking at Levaggi Ranch in Plymouth.
➤ From left: Margaret, Joan, Evelyn, and mother, Elizabeth Penrose, with the car that brought them out to California from Michigan.

HISTORIC PHOTOS COURTESY EVELYN CUNEO

© CAROLYN FOX

to us and everyone bought their cars from him."

While still going steady, Evelyn tried her luck at horseback riding. "We were driving cattle down out of Pine Grove and at the end of the 30-mile ride I got off that horse and could hardly walk!" She went back to what she knew best—driving the pickup, cooking for the cowboys, hauling animals to the stockyard, and making cheese and butter.

By 1950 they had three kids—Loree, Jim-

mie, and Verna. Kathleen was on the way and a surprise was still in the future. That year, Vernon quit his job at Pacific Gas & Electric and started raising cattle and working as a tree faller and scaler.

The house where Evelyn lives is on the family ranch—the third structure built on the same foundation. One burned to the ground; one fell down. The basement houses the water well along with two huge, oak wine vats. Vernon and Evelyn's house had to be rebuilt around them. Now and then Vernon would go down and have a sip of his homemade wine. Once he left the spigot turned to the open position. The water from the well was soon pink, and stayed pink for weeks. Evelyn was a teetotaler and not pleased. Vernon was also displeased because of all the wine that was wasted.

In her 40s, Evelyn went to the doctor for a hysterectomy. "Could you wait a few months, Mrs. Cuneo? You're pregnant!" Dr. Lynch delivered the surprise, last daughter Robin.

Evelyn celebrated her 94th birthday in December. The day after Christmas a crowd gathers for a Cuneo tradition known as "the butchering." This chilly time of year is great for processing pork and hanging the meat. Son Jimmie says, "There's a big copper bucket outside that now holds boots, but it was used to cook leftover grain or potatoes to feed the hogs."

When Evelyn's daughter Loree married Doug Joses from across the river, things started to mushroom at slaughter time. Ranchers brought their hogs to the Cuneos' place for processing. But things had to change. One year Evelyn and brother-in-law Elton tried to make Italian blood sausage in her kitchen. Soon folks from outdoors came into the kitchen to warm themselves.

"I went to bed when the wine started flowing," Evelyn explains. "Next day I looked around and the kitchen was clean, but I looked up and found blood on the ceiling." So it was time to move the hog processing to the barns. "It's fun," she says, "if you don't live here!"—*Carolyn Fox*

Tony Holt, 88

Never pick up and run.

Growing up in the middle of 14 kids, Tony Holt knew what it meant to make do. "We were hungry when we went to bed and when we woke up," he says. "That's how it was. I thought nothing of it."

The kids slept upstairs—boys on one side of the room, girls on the other. "Once in a while someone would bring a box of clothes from town. We would wear it whether it fit or not." A few years ago he realized he didn't have to wear boots that curled his toes under. "It took a while for me to buy boots that were big enough."

The kids made their own fun. "I would get inside this old truck tire and then the others would roll it down the hill. We had a great time."

Tony's father, Jack, the foreman of the coal docks in nearby Hettinger, N.D., walked to work and home again carrying a large sack of coal—part of his wage. The family raised crops and livestock, doing all the field work, haying and feeding with horses. They walked or rode to town even when everyone else was driving cars. "I had a good-looking black mare that I would ride into town just so I could look at her reflection in the windows on Main Street," Tony says, smiling.

When he was a teenager, he trailed 10 horses back from Holland Center, S.D., where his brother was working. "They were tied hard and fast, head to tail behind me. The river was flooded. An old bachelor told me, 'Don't fight your horse, let him have his head. He'll feel like he's not moving but he's moving.' I'm glad he was there. I hadn't swum a river like that before."

After graduating from eighth grade, Tony worked in a produce house that bought chickens, turkeys and eggs. They dressed the poultry and candled the eggs, then sent carloads to Chicago. Later he worked for the railroad building water tanks in North Dakota, Iowa, Montana and Minnesota.

"When I was 26, I decided I would come home, buy a cow, and follow her around until I starved to death, and I'm doing a pretty good job," he jokes. Tony moved to Ralph, S.D., to deliver mail. "I took the job for 30 days and I'm still here." Soon he was

> **"When I was 26, I decided I would come home, buy a cow, and follow her around until I starved to death, and I'm doing a pretty good job," he jokes. He moved to Ralph, S.D., to deliver mail. "I took the job for 30 days and I'm still here."**

PHOTOS COURTESY HOLT FAMILY

> **"We were hungry when we went to bed and when we woke up," he says. "That's how it was. I thought nothing of it."**

working for a local rancher when he wasn't on the mail route.

The winter of 1949 was tough. Tony delivered mail horseback when his Jeep couldn't get through. "My horses were in good shape. They had to lunge through the drifts. Neighbors offered me horses, but theirs were tender. I was better off with my own tired horse than one of theirs."

One nice day his brother rode along. "The temperature dropped 70 degrees in two hours and a blizzard hit. We got stuck in a snowdrift and walked to the neighbor's house. Ben was wearing a light buckskin jacket. He sat on their furnace for two hours before he could feel the heat. If we'd walked any further I don't think he'd have made it." They stayed for four days until the storm let up.

Tony gradually built up a herd of cattle and horses. "I like to calve, work cattle, and be with the neighbors. And I enjoyed breaking them horses. I guess I thought I'd live forever, that they'd never hurt me."

He married a local schoolteacher, Dorothy Olson, and they raised six children in a two-bedroom house. "I didn't have nothin'. I've been married for 62 years and I've got half of it left," he teases. Eventually Tony and Dorothy bought an old rancher's cows and leased the place. Later they bought a neighboring place and

CLOCKWISE FROM LEFT: Birdy, Tony and Topsy after a 45-mile, eight-hour mail delivery during the bad winter of 1949. ▶ Dorothy in 1954. ▶ Tony as a young man in 1954. ▶ Tony and Dorothy in their home in spring 2012.

moved three miles south where they continue to ranch.

"My dad would say, 'It doesn't matter how much you make, it's how much you save that counts.' Dorothy and I never looked over the fence to see what the neighbors had to try to match it. We were born poor enough to know we needed to pay for what we got. There were a lot of years that we thought it might be our last year doing this, but we didn't pick up and run. We don't believe in that."

—*Carrie Stadheim*

The Staff of Life

Nothing was wasted. No money changed hands, as the miller took his pay in wheat.
By Marge Bennett

It was spring, and once again we were plowing the rich, loamy soil, working the multitoothed harrow, leveling and breaking the clods into miniscule grains of soft, warm earth. I walked barefooted behind my dad, each step revealing my high arches and leaving little round toe prints in the freshly worked field. The planter, drawn by old Duke and Dolly, the venerable team of horses, dropped precious seeds of wheat into the furrow, then quickly covered them to just the right depth. The good earth, surrounding each kernel, cradled it lightly until it sprouted. Soon the small blades of wheat would turn the dark field into a sea of emerald green.

Growing quickly, reaching for the warmth of the Colorado sun, from within each stalk of wheat a miracle happened. One slender pedicle would shoot up. Within a few days it would begin to form the head, composed of many grains of wheat, each one wrapped in its own snug covering. In a few weeks, each stalk would ripen in the summer sun and the entire field, pregnant with promise, would take on a golden hue, waving in the breeze, waiting for the harvest.

Neighbors, helping each other, worked feverishly, going from farm to farm to harvest the plump, smooth kernels that had been threshed by the huge, noisy, mechanical monster, the threshing machine. Burlap sacks of wheat, each weighing nearly 100 pounds, were sewn with a sharp, curved needle threaded with strong twine and knotted tightly. When the granary and the barn were filled to overflowing, we relinquished even our bedrooms to sacks of grain.

The finest wheat was saved for the next year's seed, but most was loaded on the old Dodge pickup truck, and with the springs

© JOHN BARDWELL

creaking and groaning under the weight of it, Daddy would drive the short distance to the mill. He would place an order for white flour, whole-wheat flour, cracked wheat, which made a delicious breakfast cereal if soaked and cooked slowly, and "shorts" or "leavings," which fed the chickens and the hogs. Nothing

When the granary and the barn were filled to overflowing, we relinquished even our bedrooms to sacks of grain.

was wasted. No money changed hands, as the miller took his pay in wheat. The hard times of the '30s taught us how to survive.

Mama was the magician in the kitchen. She could turn ordinary flour into the most delectable goodies ever tasted. We children could hardly wait, with the aroma of fresh-baked bread beckoning us to our humble kitchen. Home from school and hungry, we would lift the snowy white flour sacks Mama had spread to reveal six mouthwatering loaves of golden crusty bread, a pan of cinnamon rolls and dozens of dinner rolls. The covered glass butter dish, filled with freshly churned sweet butter, made us feel as though we had found the "pot of gold" at the end of the rainbow. Mama would cut thick slices of the still-warm bread for each of us. A jar of strawberry jam or sweet, spicy apple butter made an even greater treat. Served with a glass of cold milk, what could be better?

It sustained, comforted, and supported us. No wonder we became addicted to "the staff of life." ■

Making Biscuits

© *Larry Angier*

Dotty and big sister, Della, make biscuits with their mother, Deirdre Lefty, and her mother, Loree Cuneo Joses, at a makeshift kitchen at Cuneo Ranch in the Sierra Foothills in Amador County, California. They are helping with lunch during the annual winter gathering when several generations of local ranching families process hogs into hams and sausage for the three days after Christmas. Later, Dotty won Reserve Grand Champion Market Lamb at the 2021 Placer County Fair, much to the delight of her sisters, parents, grandparents, and the rest of her extended family.

Lincoln Gabriel, 84

I was a cowman.

Lincoln Gabriel, 84, has always been high-spirited, in more ways than one. He has been raising cattle and sheep and growing hay since he was a young boy working alongside his father and late brother in southern Oregon.

He's also raised Cain especially as a frisky boy going to school in Olene, a ranching community about 12 miles from Klamath Falls. He recalls lots of problems, explaining: "Ranch kids was a little spirited. There was a teacher by the name of Mrs. Olson who had a helluva rubber hose—I do remember that."

Lincoln's father, Frank, kept others in spirits during Prohibition by selling whiskey. "He made his own moonshine. They all did," he says. "He made white lightning to sell at the dance hall."

At the time, the Gabriel family (including his mother, Sylvia, and brother, Frank) was living on a Swan Lake homestead his father had settled after riding a tramp steamer around Cape Horn, jumping ship in South America, and eventually winding up in the Klamath Basin.

"You use what you got, and what Swan Lake had was rye so they made rye whiskey," he says. "It was a pretty hard whiskey."

Lincoln was born in Klamath Falls but raised on his family's nearby Swan Lake homestead. When he was eight, his family moved to an 18-acre island on the Lost River. They raised hay, grew potatoes, and kept a band of sheep. "We wuz as poor as church mice."

The family moved to Olene in 1938. Lincoln and his brother took over most ranching duties because their father had another job. He especially remembers the winter of 1936-1937.

"I never seen the like," he recalls of snow that was 12 to 14 feet deep with 30-foot drifts. "I had to milk those cows. They froze their teats and when you tried to milk 'em,

CLOCKWISE FROM TOP: Lincoln, #37, during his high school football days. ➤ *Lincoln today.* ➤ *Gabriel clan in the 1950s. From left: a family friend, Frank Sr., Sylvia, Ruth, and Uncle Win Bates.* ➤ *Lincoln, right, with Frank and father, Frank Sr., in the 1960s show off antlers.* ➤ *Lincoln, right, with brother Frank after a good day of fishing.*

> **"It was Prohibition and Dad made his own moonshine. They all did. He made white lightning to sell at the dance hall."**

they'd bleed. We chopped holes in the river for the sheep. Chop 'em 10 feet deep and five feet wide. We'd do that in the morning before going to school."

Lincoln was treated for an ulcer at age 16, explaining, "It was the pressure of trying to run these darn ranches." A year later he took over ranching duties—by then cattle had replaced sheep—when brother Frank

trooped off to the military.

Life eventually settled down. He married Sandi in 1973. Until Frank's death in 2000, the brothers lived side by side, Frank in the family home and Lincoln and Sandi in a manufactured home a few steps away.

Looking back, Lincoln regrets that he focused so much of his time on the ranch at the expense of participating at potlucks and dances. "I should have done more of that. I realize that now."

"He didn't have time for social stuff," Sandi says, explaining that Lincoln and Frank had 400 head of cattle and grew 500 tons of hay annually during their prime years. The brothers worked well together. Frank liked the farming and mechanical work; Lincoln thrived with the livestock. "I was a cowman," he boasts.

After Frank's death, Lincoln sold the cattle and, at Sandi's insistence, escaped the cold winters in Arizona. But during the summers, he's content to be back at the ranch, irrigating and doing other chores. He says, "I've enjoyed this ranch and life as much as any man can."—*Lee Juillerat*

Beverly Chandler Walter, 87

A real cowboy.

Beverly Chandler Walter of Lompoc, Calif., has spent her entire life working with horses and cattle, and her amazing horsemanship has impressed the best. She's an accomplished western artist and actress. In her youth, she was a rodeo and trick rider, but cowboying was always her first love.

Born one of nine children to George and Connie May Chandler on Aug. 27, 1932, Bev grew up in Tustin, Calif. "God gave me a special gift—a love and understanding of animals, especially horses," Bev says. "My grandfather, Jim Williams, raised rare big red Missouri mules, and I just inherited my love of horseflesh from him."

She was five when she told her mother she was going to be a cowboy. "I hid my riding britches in my lunch bucket, and wearing Levi's got me sent home from school more than once."

Bev would be the first to tell you, "I'm a cowboy, not a cowgirl." She believes cowgirls just ride, while "cowboying is a profession." Until a riding accident in 2000, she worked on ranches right alongside the male cowboys doing the same work: breaking and shoeing horses; gathering, branding, castrating, roping and hauling cattle and horses; driving teams; and building and mending fence.

"I was around 10 when I started helping out on the Irvine Ranch. Lem Thrall was the foreman, and he was my mentor." While working there, Bev got to ride Mrs. Irvine's world champ cutting horse, California Polly, and other impressive western mounts. "It was Pearl Harbor and the war, and all the ranches were shorthanded. That's really how I became a full-time cowboy as a youngster." Her mother bought her a saddle from Sears and Roebuck for $65, but insisted that Bev always be a lady, and never swear or drink.

By high school, Bev was running the family's Chandler Riding Academy. She was 16

when she met Montie Montana, Hollywood trick roper and rider. He hired Bev to ride in his traveling horse show. She soon learned the riding tricks and loved entertaining, especially making the children laugh. It was through Montie that the doors opened to the movie industry. During the next few years, Bev worked with Roy Rogers, Gene Autry, Slim Pickins, Dale Evans and Bette Davis.

On Nov. 19, 1977, Bev Chandler married Albert Walter, a rancher she had known from

"One time I went to work on this ranch and one guy said, 'If she's gonna go break horses, then I'm gonna go wash dishes.'"

her youth. He was much older than her, but she had always admired him. She didn't accept his proposal right away, but shortly afterwards during a telephone conversation, she said, "If you've got a question, I've got an answer."

Albert and Bev leased a 12,000-acre cattle ranch and all was well until 1981 when it was destroyed by arson. They were left with $44 and a ton of bills. They moved to the Santa Ynez Valley and Bev reentered the film indus-

try. Still today, 28 years after Albert's passing in 1991, Bev's heart is filled with sadness. "I don't think I could ever have a better marriage."

Bev's most notable movie performances are in "Of Mice and Men" and "The Last Cowboy." In 1992, Bank of America displayed a national billboard where Bev showed her true character as a cattle person—rope draped over her shoulder and donned in her

CLOCKWISE FROM TOP: In 1992, this photo of Bev Walter was used on a national billboard by Bank of America. (It didn't include the quote.) ➤ Bev Walter's grandparents, Jim and Athula Jane Williams, and her mother, Connie May, standing between them, are surrounded by her siblings in Water Valley, Mississippi, in 1899. ➤ Bev Chandler, 4, in 1936 in Tustin, Calif. ➤ Montie Montana, Hollywood trick rider and roper, with Al and Bev in 1990.

PHOTOS COURTESY WALTER FAMILY

She was five when she told her mother she was going to be a cowboy. "I hid my riding britches in my lunch bucket, and wearing Levi's got me sent home from school more than once."

worn cowboy hat. That same year, her western landscape photos hung in the Philip Morris building in New York, and many of her oil paintings were purchased by Easterners. She is honored to be included in Lynda Lanker's collection, "Tough by Nature—Cowgirls and Ranch Women of the American West."

Proud to represent the cattle industry, she says, "I love all beef, cooked any way, but barbecued is my favorite." Bev is currently working on a memoir, "I Was Born To Be A Cowboy."—*Glenda Rankin*

Flowers for Mom

© Kayla Dixon

*With an angelic smile, six-year-old Luke Dixon
brings his mama wildflowers.*

Jack Martin Anderson, 89

Tough roots,
sheltering branches.

Jack Martin Anderson was born on the Tatham homestead to McKee and Josephine Goldman Anderson on June 16, 1923. It was his mother's homestead in the Missouri Breaks because McKee lost his place trying to farm it. Josephine remembers, "McKee had already decided to plow only the required 40 acres for proving up and leave the rest of the land for pasture, as he thought God had intended."

As boys, Jack and his older brother, Bill, rode 12 miles to Lone Pine School and stayed in a dormitory. It was five miles from the nearest homestead. Jack finished five grades in four years.

"I broke my arm trying to tightrope walk the clothesline, because I'd seen pictures of acrobats," Jack says. "The school was 65 miles from town, and I was too sore to ride home horseback. The folks came with the Model A. After the cast was off, the doctor told me to carry a pail of rocks everywhere to stretch the arm back out. I didn't want to carry it, so my folks would tie the handle to my wrist with a rope. It did straighten out and got about as good as the other. I was eight or nine."

Jack attended high school in Glasgow. Once, he and Em Uphaus were swimming in the Milk River. "We saw a great big glob of brown hair floating." The hair belonged to the government trapper's oldest daughter. "We didn't say anything, just grabbed that hair, pulled her head above water, and hauled her to shore. She was mad! We saved the girl's life, but I don't think she ever got over it." Decades later Jack left a message on her answering machine, but his call wasn't returned.

One time, Jack's dad had the team and wagon out front of the house. He was getting a drink inside before going for a load of hay. Jack went out the back door firing a cap gun. "I'd been told I wasn't to do that, but I went and did it anyway." He caused a stampede. "The team went under Highway Bridge Number 2. Jumbo went around one side of the post, and Sully went around the other. They come around and bumped heads doing about 15 miles an hour. Sully always ran away after that, and Jumbo had the crazies the rest of his life. The wagon just mushroomed. I stayed out of sight just as much as I could."

After graduation on Sept. 16, 1940, Jack joined the 41st Division, Company G, 163rd Infantry Regiment. In his book, "Warrior... By Choice...By Chance," Jack writes: "I surely did not realize that when I enlisted in the Montana National Guard at Glasgow, I would spend 24 years, four months and 23 days in military service to my country. That I would survive 51 months of combat in infantry units, be wounded twice, and be a prisoner of war never entered my thoughts."

Jack served in the Pacific in World War II and the Korean War. He was shot through the head and neck in 1951 and then captured by the Chinese, but he managed to survive and escape.

Jack enjoyed almost four years of peace between wars. On Dec. 16, 1945, he married his lifelong sweetheart, Betty Jane Hallock. Together they had many adventures and raised two sons and a daughter. After Korea, Jack became a firefighter.

When Jack hears about feds and environmental groups planning to remove his family from the old homestead for a buffalo commons, he shakes his head and calls it a "boondoggle."

In 89 years, Jack has seen a lot of boondoggles come and go, but these things have endured: faith in God, love of family, respect for the land, and the courage of one man with tough roots in Montana soil, who is willing to reach out and shelter those in need.

—*Sierra Dawn Stoneberg Holt*

> **"I broke my arm trying to tightrope walk the clothesline, because I'd seen pictures of acrobats. The school was 65 miles from town, and I was too sore to ride home horseback. After the cast was off, the doctor told me to carry a pail of rocks everywhere to stretch the arm back out. My folks would tie the handle to my wrist with a rope. It did straighten out and got about as good as the other. I was eight or nine."**

PHOTOS COURTESY ANDERSON/STONEBERG FAMILY

CLOCKWISE FROM TOP: From left: Bill (Jack's son), Jack, McKee (grandson), Zora (great-grandniece), and Elizabeth Jo (daughter). They are picnicking at the site of the old Lone Pine School. ➤ *Bill with Jack on Tud, ca. 1925.* ➤ *Jack and Betty on their 52nd anniversary in 1997.* ➤ *Jack and Bill in front of the log ranch house that McKee Anderson built around 1925.*

Ridin' Flank

© Rick Raef

Top cowhands and brothers Crae (left), age seven, and Cy, age five, ride flank together while gathering cows and calves for the Duarte Livestock Company in Beatty, Oregon. Crae and Cy along with mom and dad, Nikki and Chance Millin, traveled from their ranch near Lakeview, Oregon, to put in a day's work for neighbors.

Jan Edgerton, 80

A nice, long ride.

"When you are in love, nothing bothers you. Anything and everything is possible," says Jan Edgerton, as she describes her life with her husband, Jim, as home-steaders on a plot of land adjoining the Little Susitna River near Palmer, Alaska. "Before we had a well, I would haul 10 gallons of water on a horse from the river. You put five gallons on one side of the horse and five gallons on the other side. Hauled water every day, too."

They could only afford a sod roof on the cabin they built but Jan loved how it looked when the wildflowers sprouted and flourished during the summer. What she didn't love was when squirrels dug holes in it. "Aunt Eleanor visited us once," she says. "One morning she woke up screaming. The roof leaked where a squirrel had dug a hole. Aunt Eleanor was baptized with Alaskan rain!"

The roof eventually became a sieve from the many squirrel holes. The rain then caused steam to come off the logs burning in the kitchen's woodstove and living room's potbellied stove. By that time, Jan and Jim had saved enough money to put a proper roof on their cabin.

In 1960, Jan joined other local cowgirls to start Alaska's Girls' Barrel Racing Association. "There was no official women's western riding group anywhere in the state." Almost a year later, Jan captured the title of top barrel racer in Alaska. "I rode hard for that title, but actually winning it was a dream of mine."

In 1967, Jan and Jim took their Double J brand to a 460-acre cattle ranch in Utah's Uintah Basin. Actor and director Robert Redford filmed part of "Jeremiah Johnson" at the Double J. "I was impressed that Redford could actually ride well," Jan says. "I let him ride my favorite black horse. Then he wanted to buy it. I loved that horse more than money so Redford never did get that horse."

Not long after that, breakfast-sausage-king Bob Evans purchased five Spanish-barb mustangs from one of Jan's neighbors. Evans hired Jan and another cowgirl to transport his colts to his ranch east of the Mississippi River. "Sixty miles outside of Kansas City we had a tire blowout. A cowboy stopped and helped us. No way did he want anything to do with the mustangs. My girlfriend and I unloaded and took care of those colts ourselves. We really did appreciate all of that cowboy's help, though."

Jan and Jim's last stop for ranching has been at the E Lazy Heart ranch outside of Duncan, Ariz. In 1980, Jan became active in the Arizona Cowbelles. "The Cowbelles started off as a group of women just getting together to exchange recipes and such. Now we work on saving our beef industry, raising money to send a local teen or two to college, or just all of us gals going on a nice long trail ride."—*Sharon Wysocki*

CLOCKWISE FROM TOP: Jim and Jan on pack trip to check cattle in New Mexico for Buzz Esterling. ➤ Jan with Robert Redford in Utah on location for "Jeremiah Johnson." Part of the movie was set on the Edgertons' forest lease. ➤ Bob Evans, Jan and Mary Ann Parrish taking mustangs to Ohio in 1972. ➤ Jan rides Dillon. ➤ Jan and Jim on their 60th anniversary. He's 84; she's 80.

PHOTOS COURTESY JAN EDGERTON

"One morning Aunt Eleanor woke up screaming. The roof leaked where a squirrel had dug a hole. Aunt Eleanor was baptized with Alaskan rain!"

Marion Bond, 86
The accidental activist.

"We're ranch people at heart," says Marion Bond. She stands behind the counter of Bond's, The Leather Experts, in Lodi, Calif. A pair of chaps drapes over a table next to the antique treadle Singer sewing machine that's moved from Sonoma, Calif., to Yerington, Nev., to California's Santa Ynez Valley to Lodi. "Even on our acreage in Nevada and here in Acampo, we had a small cow-calf and horse operation. We couldn't wait to get home at night."

Near the rear of the store, husband, Kenn Bond, is sanding a boot sole. Kenn and Marion have been business partners for 41 years, married for 49. They met in the Santa Ynez Valley during the 1960s when Marion ran a dude-horse string and Kenn cowboyed on the neighboring ranch.

In 1975 they left Sonoma, where they had their first full-service leather repair shop, for Yerington. "We fell in love with the people there. We were so at home. We were there for the last few good productive years." The Anaconda Copper Mine closed when prices fell as a result of copper imports from Chile. The local slaughterhouse and commercial chicken ranch closed when the EPA came in. The small town began to lose businesses. Even the doctors left. The Bonds put together a mobile repair shop and advertised their schedule on a Fallon radio station. They traveled for two years from Hawthorne to Winnemucca and towns in between—"Wow, were we busy!" says Marion—until the gas shortage.

They moved away and settled in Lodi, where they opened the 1,100-square-foot

store they currently occupy. Nevadans still send them work. Marion points at her husband with her thumb. "This guy knows what he is doing. He trained under a shoemaker from the Old Country. He does hand lasting and welting. He even restyles boot heels or makes new leather ones. He learned the right way to do things."

The right way to do things is important to Marion. She says: "I've always been very quiet. I'm not an aggressive spokesman-type

© FRED HOLDEN

> "I've always been very quiet. I'm not an aggressive spokesman-type person." That changed when Marion and Kenn received a letter that began, "Dear Well Owner."

PHOTOS COURTESY BOND FAMILY ARCHIVES

FROM TOP: Marion and Kenn today at their antique treadle Singer sewing machine at Bond's in Lodi, Calif. ➤ *Marion and Kenn's 1975 store opening in Yerington, Nev., inside and outside.* ➤ *Kenn shows Black Time at the stallion parade in Lodi in 1965.*

person." That changed when she and Kenn received a letter that began, "Dear Well Owner." The North San Joaquin Water District planned to tax private domestic wells to build infrastructure to artificially recharge the aquifer. "We put the well down, we pay the electric bill, what is going on here?" asks Marion. "It's the only time we got off our duffs and said no."

Marion and Kenn made phone calls and

stuffed fliers into mailboxes, alerting people to the tax. They wrote letters to the editor of the local newspaper. At the first public meeting on the issue, 375 people showed up at Hutchins Street Square in downtown Lodi, standing room only. When Marion walked in, she said, "My God, it's starting to pay off." To the Bonds the overriding issue is the limited capacity of our natural resources.

"Overdevelopment, changes in land use, and an ever-increasing population require wise decision making," she says. "We need government agencies working together for the good of the area, not for the developers and fast money."

Through this experience, Marion learned that she had the courage to stand up for her convictions. "You can't just sit back and expect everything to turn out alright."

She talks about the family farm: "It used to be that people could actually survive and raise a family on their own property. They could have a little fruit stand out front. Now it's all regulated. Are you kidding? We're getting food from all over the world the FDA doesn't inspect. So what is going on?"

Marion's clear blue eyes are wide. "Our country is worth fighting for. There have been enough people who gave their lives for the freedoms we have. We still have the best country in the world. But what makes us think it's all automatic? You have to keep working for it."—*Suzanne Finney*

John Hoiland, 86

All alone and happy.

At first glance, John Hoiland's rustic cattle ranch looks like a junkyard, with old, dented, rusting or decaying objects everywhere you look. In the front yard of his ramshackle old house two battered, ancient 10-foot TV dish antennas rust in the weeds. In the backyard you'll see an odd collection of ancient cars, trucks and tractors—most exactly where they stood when they died.

There are several six-foot-tall stacks of wood—not the new, clean wood that might be used for construction, but gray, weathered, ancient chunks of all sizes and widths that most people would haul to a burn pile. Next to them are other tall stacks of metal pieces. They are long, short, round and square. Some shine with chrome; others hide behind a curtain of rust. No matter what you're looking for in metal, John has a piece of it resting there in the tall grass.

Then, just when you have decided that this can't be a cattle ranch, the squeaky old door at the back of the squeaky old house swings open and out steps one of the most unique ranchers in Montana. John Hoiland, 86, wears a small smile as he spots the confused look on my face. "Newcomers just don't recognize all the good stuff around here," he laughs. "They've never seen a world-class saver before."

The history of the Hoiland ranch began in 1926 when Emanuel Hoiland, a recent immigrant from Norway, purchased the land 23 miles back on a dirt road between Livingston and Big Timber. John was born at home—"It was a snowy March and there wasn't a nearby hospital or doctor"—and has lived in the house his father built ever since. After his parents died, he learned to live there alone.

His father ran sheep on their 800 acres, but John switched over to cattle and got rid of his horse. "I don't need to chase them around on a horse. I can handle my 180 cows on foot. Herefords are such sweet

> **"Herefords are such sweet beasts that they just go wherever you point. They're so nice and gentle you could invite them in for dinner."**

PHOTOS © BILL KILEY

CLOCKWISE FROM TOP: John Hoiland with one of his old tractors. Notice all the other "saved" machines in the background, including that blue 1960 Chevy he uses to drive to town. ➤ John plays one of his 11 accordions. ➤ Some of this collector's "maybe I'll need them someday" vehicles.

beasts that they just go wherever you point. They're so nice and gentle you could invite them in for dinner."

John grew up in the midst of the Great Depression. "Anyone who grew up out here in the High Lonesome in those early '30s learned how to save things. When something broke you just fixed it. You never threw things away. Then World War II came along and because of gas rationing I couldn't get to town, so I never got to high school. I just learned how to make do."

The car he drives to market once a month is a 1960 Chevy—unless there's snow on the gravel road. Then he drives his old International Scout or an old four-wheel-drive pickup. "I'm one dang good mechanic. I can take any car apart and put it back together again and make it snort down the highway."

He also has trucks and a variety of tractors and farm machinery, "and a big old dozer to help me kick aside the snow when I have to get out." John's kitchen table is covered with newspapers, old radios and tape recorders. There are small stacks of "things" along the walls.

He has electricity in the house for the lights and refrigerator, but does his cooking and heating with an old woodstove, which contains a coil to heat his water. "I'd be dead if I wasn't a good cook. If I'm not too busy, I eat breakfast, but often just forget it. For dinner and supper I have stuff like macaroni and potatoes along with antelope, hamburger, deer or elk."

Even though no vegetables or fruit were mentioned, the diet seems to be good for him. He strides across his land cutting, chopping, building, tending and watering without missing a step. He doesn't wear glasses or hearing aids either, and has cut his own hair since his mother died.

Although he lives alone—and loves it that way—he doesn't consider himself a hermit. "I have lots of friends and visitors. Strangers come and ask if they can fish in my river or hunt."

His favorite hobby is playing accordion. "My dad paid for his trip from Norway by playing for fellow passengers. I've never had a music lesson, but he taught me how to play it, and I also learned how to handle a piano and an organ. I can play a piano, an organ, and a guitar all at the same time—although it hurts my arm to twist it like that."

John inherited his dad's accordion. And then he bought another and another. Now he has 11 accordions, which he plays anywhere folks gather for fun and music. He enjoys the instrument so much that he often sits out in his backyard on moonlit nights and plays for his cows.

He also saves old hats. The one he usually wears looks like it has been attacked by something mean and hungry. "My favorite hat blew off the day Nixon died, and I've never seen it since."—*Bill Kiley*

Cowboy Enthusiasm

© Marion Dickinson

Little cowboy Dean (in the big hat) explains branding to visitor William at a ranch near Otto, Wyoming. William is flummoxed.

Riding Shotgun

Anola is always Clevon's best hand.
By Sophie Sheppard

© LINDA DUFURRENA

Anola, 95, and young husband, Clevon, 90, in 2010.

Ruth Lake says her good friend Anola Dixon "is like a firecracker on the Fourth of July. Cute as a little girl's dolly and just as fun." It's true that Anola and husband, Clevon, are both small people, but the cuteness hides internal steel that hasn't dulled in 95 and 90 years, respectively. They are tough but with a sense of humor that is always bubbling over into laughter. And gentle, too.

Anola's life with Clevon played out in that empty corner where Nevada, Oregon and California meet. It is one of few places in the continental United States where the population has dwindled in the last 60 years to almost zero inhabitants per square mile. It is the darkest part of the country when you fly over it at night, with hardly any lights at all for several hundred miles. It wasn't all that populated when Clevon and Anola were there either.

Anola Hapgood was raised on the Last Chance Ranch, which was owned by her parents, Jesse and Olive Hapgood. The main house still stands—that old stone building in the Little Sheldon Antelope Refuge

PHOTOS COURTESY ANOLA DIXON

ABOVE: *Anola, left, age five, and little sister, Barbara, with sage hens caught in front yard at Last Chance Ranch.*

LEFT: *Anola's grandmother, Mary Jane Jones, with Anola sitting in her lap. Barbara is standing with their uncle, Mary Jane's late-born son, Raymond.*

Anola, pre-Clevon, in Denio, Nevada, in 1938. She was 15, returning to the ranch after a dance the previous night. "I had friends in the three-piece orchestra."

that U.S. Fish & Wildlife Service has shored up and reroofed so it will stand for a long time. On the hill behind it are stacked rock walls—kids' forts that Anola and her sister and brother built. Remnants of old tin kitchenware they might have played with are scattered in the sagebrush. The Last Chance was the Hapgoods' ranch until the property was sold during the Depression to pay debts incurred by Anola's grandfather. When Anola was eight and her sister, Barbara, turned six, the Hapgoods moved into Cedarville every winter so the kids could go to school.

In spite of entering school two years late, Anola was smart and liked schoolwork so much that she skipped grades and entered high school at 12. She graduated at 16 as the salutatorian of her class. She says she wasn't very social in school, preferring to spend a lot of time alone after being raised so far away from people on the Last Chance Ranch. She worked as a telephone operator for a few years until, as Ruth says, "Along came Clevon Dixon, a good-looking buckaroo and Anola snapped him up."

Anola and Clevon married in 1942. "Clevon had an old Chrysler," Anola says. "The tires were so old the inner tubes were bulging out of them. You couldn't get tires during the war. We drove it to Lakeview and then we got on the bus to Reno. We walked in big loops all over that town trying to find the courthouse. On our final loop, Clevon said we had better find it or he was going to go get back on that bus and go home. Well, we found it alright and the judge married us. The next morning we got back on the bus to Lakeview and then into the Chrysler for the last leg of the trip. On the way home, he drove it across the creek and it just stalled...right in the middle of the creek. He got out and came around and opened the door and carried me on out of there."

Her first years of married life were spent in Guano Valley on the IXL Ranch working for Ross Dollarhide. "We never saw any women out there. It was all men. God, that was awful. When we left after four years, I said to myself I would be happy if I never saw that place again." They lived in the stone main house. In the summer, the hired men bunked in the long wooden addition to the main house, and in the winter, Clevon and Anola had the place to themselves while they managed the cattle that wintered there.

"There was a big fireplace in the stone house but it didn't work very well," Anola says. "I have never been as cold as I was in that house. All those rock walls would never heat up. One time I had some carrots cut up in the kitchen. I put them in a bowl of water so they wouldn't dry out 'til I was ready to use them. The next time I looked at them, the water had iced up."

For the young woman who was an accomplished piano player, life in the cow camps must have been hard. But she didn't complain. There was always a story, and her brown eyes crinkle up with her laughter in the telling. In Guano Valley, their only neighbor was an old Irishman

Clevon Dixon in Coleman, Nevada, in 1946. A great horseman and buckaroo, he was longtime cowboss of Oregon's legendary MC Ranch.

who lived 12 miles north. He had a car and made regular trips into Lakeview and would bring them eggs from his chickens and their mail from town.

That first winter of Anola's married life was the toughest. In February they were snowed in and ran out of food. No flour. No sugar. No coffee. No potatoes. Only beef to eat. Anola told Clevon, "We've got to get to town and get supplies."

So Clevon saddled a young horse for himself and Anola's steady horse for her, and led a packhorse through the two feet of snow and deeper drifts 18 miles to Coleman Valley. They spent the night with the Coleman winter cow-camp buckaroos and Anola said the bed they slept in that night was so grimy she didn't want to sleep in it. She put one of her jumpers between her head and the pillow and lay on an extra shirt. She didn't sleep too well but was ready in the morning for the car trip into Adel where they got supplies. Another night spent in Coleman and then they loaded up the packhorse and headed home. The snow had been soft two days before, but going home the horses had an awful time of it, breaking and wallowing through hard-frozen crust.

Clevon got tired of leading the packhorse as the horses lurched through that deep snow. He unhitched the lead rope, saying, "It will follow us home." Clevon was wrong. "That horse turned right around in all that snow and headed back toward Coleman," Anola says. "There went all our food, bouncing away in the packs on that horse heading out across the valley." Clevon had a tough time getting his young horse to circle out in front of the runaway in deep snow, but he finally caught it and didn't let go of that lead rope again. At times Anola couldn't see Clevon's horse even though it was right in front of hers.

Clevon says that Anola was his best hand. Anola says she was scared of horses and called them beasts, but she rode whenever she was needed and drove the teams for the hay wagon and Jackson fork. Clevon looks at her with pride and says, "She sure was good with that team and that cart. She drove and I forked nine tons of loose hay to the cattle per day

ABOVE: Anola, left, friend Wilma Ray, center, and sister, Barbara, at Anola's high school graduation in Cedarville, California, 1932.
RIGHT: In summer, the Jackson fork could move a big wad of loose hay via a derrick and boom and swing it up to the top of a loose haystack. It had trip wires to both pick up and release the hay. It took a team of strong horses to run that operation and another team to pull the hay wagon. In the winter, the operation was reversed, using the Jackson fork to lift hay off the stack and into the feed wagon. A wagon would hold more than a ton, but two little people managed this by themselves in rough conditions and rougher weather for months at a time. Anola says they were usually finished by noon every day. She drove while Clevon forked the hay to the livestock. Then they went back to the stack where Anola would drive the other team to load the hay while Clevon ran the fork. They were a perfect team! They fed nine tons of hay a day, which is a pretty phenomenal daily feat to us today, but was normal for ranching in the '40s.

that winter."

Anola may be afraid of horses and dogs, but not snakes. "It's a good thing because there were so many of them around those camps," she says. "One time I was going out with my paring knife and a bowl to get vegetables for dinner out of my little garden at the IXL. All I could grow was carrots and onions and potatoes because it was so cold. Stretched out across the garden path was a big, old rattlesnake. I stabbed it with my paring knife real quick. It was still there a couple of hours later. Couldn't go anywhere with my knife stuck through it clear into the ground."

Another time, Clevon tells that she caught a rattlesnake by the tail while they were walking through some big sage. She carried it for a few minutes until Clevon dangled a dead mouse in front of her that he had found on the same trail. "She screeched and threw that rattlesnake off into the brush," he chuckles. "She is more afraid of a danged mouse than she is of a snake."

About 10 years ago, I was on one of my daily walks up Lake City Canyon with my old dog Pete. On the way back, I met an agitated Clevon who had dropped a big, dead pine snag and it blocked the road. He was in a hurry to get home, fire up his tractor, hook up the wagon, buck all that wood up, and get it loaded and off the road.

We had friends staying with us that weekend, so I rousted them and told them we would get a little early morning exercise. We could already hear Clevon's Johnny-popper heading up the canyon. Lynn grabbed his chainsaw and in about 10 minutes we had helped Clevon and Anola cut the tree up and load it on the old hay wagon. Clevon thanked us. Anola climbed onto the tractor and stood behind him, one foot on either side of the axle housing, her hands clasping his shoulders. I looked up at those two small people, the ancient tractor and rickety wagon loaded with firewood and was worried. "Anola," I said, because she was about 85 at the time, "do you think that is really safe?"

"Oh," she said, "he had better not dump me off and run over me. If he does, I'll never let him hear the end of it." And off they went. ∎

Sophie Sheppard is a painter, rancher and writer from Lake City, California.

George Parman, 81

My heart was always in ranching.

George Lawrence Parman was born in Lake City, Calif., in the cold winter of late 1931. His father, Lawrence, was a fourth-generation rancher. His mother, Jewelle, was a rancher and an accomplished artist. George had an older brother, Joe, and sister, Christine. He says: "My dad and his brothers had properties from Lake City in Surprise Valley, California, to the Soldier Meadows Ranch in Nevada. Dad was born in 1897, and '31 was hard times and the Great Depression eventually did cost them their properties. It cost hard feelings and tears."

© ALAN HART

The family ended up in Carson Valley, south of Reno, where Lawrence milked cows, trapped coyotes and anything else he could. Lawrence's big passion was horses, which he passed on to George. Lawrence started many horses for the ranches in Carson Valley. George's mother "was an experienced trapper and those sheepmen over there needed trappers bad."

An ad in a newspaper about a commissioned livestock sale in Reno played an important part in the family's life. "Lots more than Dad ever realized at that time. The auction started a multimillion-dollar livestock business."

One of the ranches was the Smith Creek Ranch. World War II started and it was impossible to hire anybody. "It worked us to death," George says. "We wound up with the Weeks Ranch and Dad couldn't see much sense of school, so I quit school in the seventh grade." But George could see the value of an education, so he went back after a couple of years and rode with Ruthie Weaver, who was born in Minatare, Neb. She had a ranching background and came to Nevada when her father got a job in Hawthorne. George, with help, caught up and graduated high school in 1951. Ruthie and George were married that year. He was 20; she was 17.

FROM TOP LEFT: George Parman with bobcat skins in Diamond Valley, March 2001. ➤*George and Ruthie with children, Linda (on left) and Georgie, in Hunt's Canyon, August 1955.* ➤*The cowboy in Eureka in 2012.* ➤*Mustangs caught on Table Mountain in Bald Mountain Corral, ca. 1965.* ➤*George's father, Lawrence, with coyote and raccoon skins in 1934.* ➤*Georgie and Linda on their way to school in a snowy March 1964.*

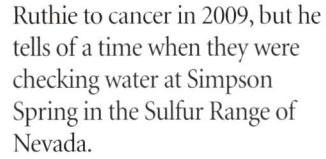

They had three children: Linda, Georgie and Robin.

George and Ruthie moved to Hunt's Canyon in Monitor Valley in 1954. "I traded around. I loved to trade. Ruthie had a fear not to borrow money. We had lots of cattle. We inherited a range conflict. We were strange to the country, but we were forced to ride—I'm proud of that."

The drought hit in '56-'57 "and prit' near put all of us out of business. We got through with a few cattle." The drought ended in '63 and they paid for the ranch by selling cattle. "We didn't owe a dime," George says. "We sold out and ended up at the Flynn Ranch in Diamond Valley. Ranching's hard...damned hard. And I went through some pretty rough years. I don't know how Ruthie ever stood me all those years. When I got too much money, I always spent it running horses."

After 58 years of marriage, George lost Ruthie to cancer in 2009, but he tells of a time when they were checking water at Simpson Spring in the Sulfur Range of Nevada.

"We came up to this spring pipe that was plugged up. Ruthie took off her boots and socks and waded into the tank which had about two feet of water. She was unplugging the pipe and I said, 'Watch out for that snake.' Ruthie scolded me saying, 'Here I am wading in this cold water and you are never serious.' Just then she spied a small water snake trying to climb up her leg. Ruthie then tried to climb a tree that wasn't there. She sure got mad at me."

George still misses her. "Ruthie and I might have dodged some of the pitfalls we had through life, and I think the life we had was kinda hard at times, but we'da probably done it much the same way. Ranchin' was a hard ol' life but we both loved it."

—*Paul J. Etzler*

John, 91, & Tansy Smith, 89

The Owens Valley farmer and the cattle driving queen.

Finding time to visit with John and Tansy Smith isn't always easy. Depending on the season, John will likely be operating farm equipment, raking or cutting hay, or doing chores on their ranch, extending between Independence and Manzanar on the east side of the Sierra Nevada.

On any given day, Tansy might be out riding horseback. Two days before my visit, she had been moving 200 head of cattle, and a day earlier she was searching for stragglers. "I rode from dawn to dusk," she says matter-of-factly. For John, who's 91, and Tansy, who turns 90 in December, working long, hard hours is part of their routine.

They run a 300 head cow-calf operation on 5,000 acres of brush. They irrigate 500 of those acres for alfalfa and some orchard hay. Each fall, the calves are sold to the Harris Ranch. Until 1982, when they switched to a cattle operation, they operated the Independence Dairy.

John was born and raised in Utah and spent his high-school years in Kansas. He met Mary Fitz-Patrick, known as Tansy, in Ventura, Calif., after he and his father moved west. As a youth, Tansy helped her mother and stepfather, Archie Dean, run a pack train, and she milked cows and fed calves on their family ranch before going to school. She rodeoed and competed in horse shows in pole bending and jumping. And she was 1941's Miss Ventura County.

"Tansy is really the cattle person," John says. "She was, and still is, in charge of the cattle operation. It was hard for me because I had to do my share of ranching and farming after work."

A 1946 Stanford graduate (his studies were interrupted during a three-year Navy stint during World War II), John went to work for Inyo County after college. He and Tansy had married in 1944 in Florida during his Navy years. After moving to Independence, the Inyo County seat, they bought the dairy, where they still live, in 1948. John held a series of jobs with the county for 33 years and served as county administrator until he retired in 1981. He recalls working 10- to 12-hour days at the courthouse. "Then I'd come home and have to go out and milk the cows—unless my wife or kids helped out."

Tansy stayed at home, raising their four children. She remembers riding horseback with one child in front and another behind. She was also in charge of the dairy and, later, the family's cattle operation. They produced and sold raw and pasteurized milk and cream

© LEE JUILLERAT

CLOCKWISE FROM TOP LEFT: John and Tansy still live on the ranch they bought in 1948, which used to be a dairy. ➤Ensign Smith and Tansy during the war years. ➤Moving cows in Owens Valley with the Sierra Nevada range providing the backdrop. ➤John often calls Tansy "The Queen." Here she is as Miss Ventura County in 1941.

They produced and sold raw and pasteurized milk and cream during the dairy years and she delivered milk to homes in the morning and to the jail and grocery stores in the evening. "If we retired," John says, "I don't know what we would do. I have to have something to do every day or I'd go nuts."

during the dairy years and she delivered milk to homes in the morning and to jail and grocery stores in the evening. The couple also raised, showed and sold registered quarter horses.

"We had to work hard, by golly," John says. But "when you're in your 20s, you don't mind working like that." Their family ranching tradition is being carried on by son, Zachary, who operates a large alfalfa ranch in nearby Lone Pine.

John and Tansy have no plans to slow down. Because of surgeries, John no longer rides, but he stays involved in day-to-day ranch operations. "If we retired, I don't know what we would do," he says. "I have to have something to do every day or I'd go nuts." Tansy adds, "I'm just glad to see the dawn break every morning." *—Lee Juillerat*

Take It Slow

© Rick Raef

Six-year-old Ty Duarte pushes cows and calves for a neighbor with a steady hand during spring branding on the River Springs Ranch in Bly, Oregon. Ty, sister, Madison, and parents, Eric and Nikki Duarte, run cattle on their own ranch in nearby Beatty. The family also owns and operates Duarte Sales, a local auction service company.

Jay Nelson, 90

Valley of 10,000 haystacks.

Jay Nelson's paternal grandfather, Soren Nelson of Denmark, walked into the Big Hole Valley of southwestern Montana in 1889 with a nickel in his pocket. As soon as he became a citizen in 1896, he filed on his homestead, bought another from his brother, married Lena Hirschy, the daughter of Big Hole Swiss-cheesemakers and ranchers, and never looked back.

"The family brand E-8 was registered in June of 1895," Jay says. "Granddad raised Hereford cattle and once shipped steers to Alaska with his father-in-law."

The beaverslide stacker was invented in the Big Hole in 1910 to help put the native grass hay into 20-ton stacks. "My brother Tom and I built over 50 of these in the 1950s and a few are still in use, but gradually the modern round balers are changing what's called The Valley of 10,000 Haystacks." Jay also built seven scale-model beaverslides and devoted more than 20 years at July's Bannack Days showing the tourists how they work.

"We had another unique Big Hole winter glider, the snoplane, that was used in winter for many years and races were held to see who was fastest. It had an airplane motor and three skis and my dad had two of them on the ranch."

Jay and his brothers rode horseback to school in Jackson—about three miles as the crow flies. "There was no real road to our ranch so we cut across the fields. I was told I had to go to school to learn how to read, but after my first day at school I came home and picked up the paper and still couldn't read! We trapped along the way so were always busy."

In 1941, Jay married Jean Renz from Dillon and they raised six children. When World War II came along, they had one daughter but in June 1943, "due to a misguided patriotic idea," he joined the Army just short of his 20th birthday.

"I was in Germany and had crossed the Ludendorf Bridge at Remagen in the middle of March 1945, and was sent to take my machine gun out to the end of a ridge and see that nobody came behind us. Nobody told me there was a large Tiger tank coming up a timber road until it was right below me. With a monumental stroke of poor judgment and pure stupidity, I decided to put a

> **"I was told I had to go to school to learn how to read, but after my first day at school I came home and picked up the paper and still couldn't read!"**

few rounds through the open turret. That upset that German soldier to no end and he proceeded to shoot the end off the hill that I was on. He did turn around and go back but my war ended right there."

Jay was taken to Linz to a field hospital in a large schoolhouse for almost a week. From there he was sent to a Brit hospital in Southampton. After time in Fort Sam Houston he was discharged from the Army in November and returned to the Big Hole.

"The Big Hole is famous for two big events," he says. "Clark of the Lewis and Clark Expedition stopped by the boiling hot springs to cook some meat on their return trip. The hot springs still provide water to the residents of the town today and are used as a swimming pool for vacationers at the Jackson Hot Springs Lodge." The 1877 Battle of the Big Hole between the Nez Percé Indians and U.S. soldiers is some miles from the Nelson ranch. "Some of the Indians stopped on the third day after the battle. We have found some spent shells on the banks of the creek that runs through the ranch."

Jay served 41 years as a director for the Vigilante Rural Electric Association in Dillon. He belongs to the VFW and has served on the school board, sewer and water board, and fire department.

Jean passed away in 1997. Jay still lives in the home he built in 1950. Family reunions

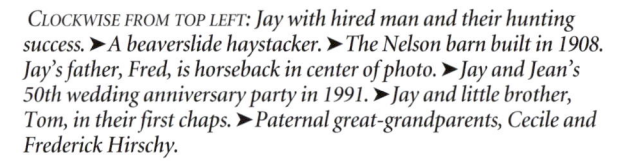

CLOCKWISE FROM TOP LEFT: Jay with hired man and their hunting success. ➤ A beaverslide haystacker. ➤ The Nelson barn built in 1908. Jay's father, Fred, is horseback in center of photo. ➤ Jay and Jean's 50th wedding anniversary party in 1991. ➤ Jay and little brother, Tom, in their first chaps. ➤ Paternal great-grandparents, Cecile and Frederick Hirschy.

are held every five years with a raffle of homemade items to fund the next one. Jay has written two books about his life in the Big Hole: "A Scrapbook From the Big Hole" and "Big Hole Memories."

—*Ruth Little (daughter)*

PHOTOS COURTESY NELSON FAMILY

Bill Ruehle, 83

The Freeway Cow Catcher.

Bill Ruehle has been a wild-cow catcher, wild-horse catcher, roper, bull rider, bronc rider, rodeo pick-up man, Army sergeant in the Korean War, cattle buyer, horse trader, mule-train packer, wheat farmer, rancher and cattle exporter to China. He supplied eight-up horse teams and wagons to the movies and modeled for the famous artist, Lon Smith.

His father, Benjamin, roped for Teddy Roosevelt in the 1900s in North Dakota. "When I was nine," Bill says, "I drove workhorse teams for 11 cents a day." In 1946, at age 16, Bill joined the Los Angeles Union Stock Yards Roundup Association. "It was there I learned to rope mean, mad, wild cattle that fed on locoweed. They would charge after me like the wild cattle out around Death Valley."

His citizens-band radio call sign is Cow Catcher. That's what the cattle-truck drivers, police and Highway Patrol called when they wrecked or lost cattle on the freeways going to stockyards or meat-packing plants in Los Angeles. "There were 4,000 cattle a day in cattle trucks on the freeways with three or four wrecks a year and a half-dozen stray cattle lost on a weekly basis usually in the middle of the night." He developed a keen sense of roping in the dark with no misses.

Bill's right thumb is stiff from a machine gun in the war so he did not dally his 35-foot rope. He tied it hard and fast to the saddle horn, even when roping big bulls, which he would usually rope forefoot, trip, then hobble.

He was born in Youngstown, Ohio, on Aug. 26, 1930, the youngest of nine children. His father had a ranch in Bixby, S.D. The family moved to Los Angeles in 1943, and in the '50s and '60s Bill rounded up wild horses in Reese River Valley near Austin, Nev., and shipped them in large cattle trucks to Los Angeles. He tithed 10 percent of his earnings to a church or orphanage and believes he had divine intervention of adrenaline to place his rope around the neck of steers in the dark of night with no misses. "If I did not tithe, then there were no cattle to rope and no horse buyers until I did." His wife,

PHOTOS COURTESY BILL RUEHLE

CLOCKWISE FROM TOP: Bill Ruehle, the Freeway Cow Catcher. ➤ Bill looks happy with his day's work. ➤ A house full of memorabilia. The Los Angeles Times took hundreds of photos of Bill. ➤ This is another successful freeway catch.

Roni, drove the pickup and trailer, and operated the trailer gates.

Bill and Roni's son, Robbie, is a church pastor, construction contractor, and roper, and their daughter, Lori, is a hospital unit secretary, ranch roundup roper, former rodeo queen, Miss California Beauty, and Bob Hope Classic Girl. There are four grandchildren.

A lifetime VFW member commanding the rifle squad and honor guard for a number of years, Bill still hauls a few horses and cows for people in his town of Beaumont. "I am a Republican conservative and Tea Party Patriot and I write many letters to editors."

He admits to eating chicken once every 10 years, but never eats fish—which he had to eat in China. He is six feet, two inches and 180 pounds, with a 34-inch waistline. "I have good blood pressure and no cholesterol problems." He eats bacon, eggs and potatoes

"I never drank alcohol, beer or soda. But I like five or six cups of coffee a day. No salad!"

for breakfast at six every morning, a roast-beef sandwich for lunch, and steak or ground beef and vegetables for supper. "No salad," he says. "I never drank alcohol, beer or soda. But I like five or six cups of coffee a day."—*Austin Rece Hough (grandson)*

In Spite of Fear

Dean Gilbert had no choice, so he did the work allotted on what he thought was the dangerous outback of Montana. By Bill Kiley

Dean's grandparents, Cora and Dudley Gilbert, on their wedding day in 1902. Cora walked from St. Louis to Montana behind a covered wagon. BELOW: *Dean and Jo on their home range.*

Dean Gilbert, age seven, and his nine-year-old brother, Paul, went out alone to tend sheep in wide-open Montana. "Flat broke in the middle of the Great Depression, our parents did what they had to do to keep us all alive and together. If that meant sending two little boys miles away to care for sheep, so be it."

Dean never met his grandmother, but she filled his dad with stories of walking behind a wagon train all the way from St. Louis in 1879. "She told Dad that only the very young or the very old got to ride in the wagons. Her main job was to pick up buffalo chips so they'd have fuel for the cooking fire at the end of each day."

Dean's dad, Lester, knew that hard times had long been with his family, and he tried to figure out how to start a sheep farm.

"It took time and a lot of hard work," Dean recalls. "He got started by accepting free bum lambs from surrounding sheepmen. People shake their heads when they hear about how two little boys were sent out alone to herd sheep, but we didn't see anything wrong with it. That's just the way things were back then. You worked hard and long for danged near nothing. Paul and I often took a few hundred sheep from local pastures and moved them seven miles away, close to the old abandoned cabin that my grandparents had lived in when they first came here."

Their mother, Ruth, would pack sandwiches, a cake, fruit and canned goods to hold them over until either she or Lester could get up to the old cabin to bring them supplies.

"We cooked," laughs Dean. "We'd mix some flour in water and try to start a fire so we could have hotcakes or fried potatoes, or we'd open a can of beans for dinner. Sometimes we just ate stale bread or oatmeal all day. If we ran out of food we'd take our fishing poles down to a creek and catch some supper. I have to admit that we were lonesome and scared a lot."

They were in bear and mountain lion country, so they were always looking over their shoulders. "Of course, there was no electricity in that old cabin—not at home either, but there we had lanterns—so we huddled together in the dark, imagining all sorts of creatures sneaking up on us. But we lived through it, and came out healthy and happy." The sight that scared them most was the appearance of a passing human. "About once a month we'd see a rider far off in the distance, or catch sight of a faraway fence mender. They made us very nervous, and we'd hide until they were gone."

Dean laughs as he recalls adventures moving the sheep. "Sometimes, when you had to move the sheep for even a single mile, you'd walk back and forth for 10 miles in order to keep them together, so my father made us a tin dog. He put a bunch of old tin cans onto a loop of wire, and that's how we controlled the sheep. We'd just shake that old tin dog and they'd get back in line."

The sheep usually did what they were

"Let's go home," she said.

told, but there were times when human intervention was necessary. "They'd come to a little stream only a foot or two wide, but they'd refuse to cross it. So Paul and I had to shove or toss all those sheep across the water."

Eventually things got a little better and the boys were allowed to go to school. "Here, again, we were often on our own. Dad rented an old cabin near the school and we stayed there, caring for ourselves for a long time. We didn't know how to play, or how to waste time, or how to just visit with other kids. It was a long time before I learned to enjoy company."

After they finished high school, their dad decided to retire. In those days everything went to the oldest son. "Paul was the oldest, so he bought the ranch and everything on it. I had to leave and try something else."

For a few years Dean tried to start his own sheep ranch, but he soon realized that he was making a lot less money ranching than he could at a job in town. So the boy who had nothing but ranching in his blood began decades of working for the Forest Service, for railroads and phone companies. In 1969 he met and married his wife, Jo. They had two sons, Daniel and Judson.

Before Dean retired in 2001, the family lived in 23 different homes. Jo had earned her teaching credentials and had been saving for that moment. She said, "Let's go home." She meant *his* home—back in the rolling hills of southwest Montana. Eight miles from the nearest small town, up a dirt and gravel road, three miles from their closest neighbor. They bought land and built a home only a mile from where his parents and grandparents had worked their ranches, in the same hills where two little boys had become expert herders.

Now 74, Dean says: "We have a small herd of cows now, not sheep. Paul's son Kenny brings a few hundred of his cows over for the grass, and in many ways, it's like old days again." Neither of Dean and Jo's sons has any desire to follow in his footsteps on a ranch.

"But I did," laughs Jo. "When I first met Dean, I knew in my heart that he was a born rancher, and over the years I knew that he'd be happiest if he could get back to it. I love cowboys. When I was little my heroes were always cowboys—and I finally got one of my own!"

At the end of the day, Dean says life taught him well. "Having someone to love is the most important thing of all." ∎

Bill Kiley lives in Livingston, Montana.

FROM TOP: *Trio on right with dog is Lester Gilbert with sons Paul (left) and Dean. Group on left is Lester's brother, Ford, with Joan and Jean, half-sisters to Ford and Lester.* ➤ *Dean's parents, Lester and Ruth Gilbert, on their wedding day on June 7, 1935.* ➤ *Dean's father, Lester, supplemented the family's ranch income by playing saxophone in local bands in bars and hotels in Livingston, Bozeman and Yellowstone National Park. He made $20 a night for his musical outings and still looked like a cowboy, sort of.*

Sadie Renfro, 90
Thirty-eight miles of dirt.

The dirt roads went on and on in New Mexico. Sadie Brown didn't care for them. During school years, she lodged with town families far from her own on the remote Brown Ranch in Long Canyon, 30 miles northeast of Folsom. "Winters were hell," she says, "and summers at home were heaven."

She spent her first four school years across the border in Branson, Colo., 20 miles from home. Then she spent six at Folsom School, and another year with an aunt and uncle at Otto, near Clayton where the Dust Bowl had recently changed everything. Just before her 17th birthday, she graduated with the class of 1940 from Des Moines High School, 38 miles of dirt road from home.

"My grandfather, Texas cowboy John Thomas Brown, moved to Long Canyon and built the rock ranch house in the mid-1880s, where I grew up," Sadie says. Today she lives a quarter mile up the road and her daughter Margaret lives in the rock house with husband, Danny. Nephew Henry Brown lives a quarter mile in the other direction. For 130 years, this valley has sustained generations of Browns.

Sadie attended business school in Albuquerque and in 1941 was working at Jordan's Clothing Store. That's where she met Kenneth Renfro. "He'd ride by on his motorcycle and whistle at me. I'd whistle back." They married in February 1942. They lived in the city and raised two daughters, Gloria and Margaret, but they all enjoyed extended stays at the ranch.

Sadie was vice president of a bank before retiring in 1980 to move with Ken back home to the Brown Ranch. They set their house beyond the rock house where they could see the whole valley. Together, they built up the cattle operation, winning awards for their conservation and grazing management practices. Since Ken's passing in 1999, Sadie has continued with the help of her nephew and, more recently, daughter and son-in-law Margaret and Danny O'Quinn.

"I had the most wonderful childhood out here," Sadie says from her lush lawn. "We raised red shorthorns. Mom raised chickens and the garden. She canned, and we traded butter and eggs for groceries. Dad would hire hands to help with the hay crop. Mom and I cooked for the crews. We raised hay with just rain and flooding, opening the

At home in Long Canyon, Sadie takes her evening walk. Deer, elk, and wild turkeys share the trails. Bears and rattlesnakes are more common than she'd like.

© TIM KELLER

PHOTOS COURTESY SADIE RENFRO

CLOCKWISE FROM TOP: At 90, Sadie still loves her ranch life. ➤ With husband, Kenneth, in 1957. ➤ First birthday in 1924. ➤ Sadie is a passionate gardener.

dams, and we used the team and wagon."

Last summer, the only green was Sadie's lawn. "Since 2000, it's been heck here, it's been so dry." She feeds her 50 cow-calf pairs daily. "I go out with Henry once a week to check them all. He has his cows and I have mine but we buy together and ship together. I order my own hay, do all my own business management. I've never missed a branding or shipping. I've always been proud of my cattle. They're my life."

At 90, she's still pretty independent. She does her own grocery shopping and cooking and is a passionate gardener. Last year she canned a bumper crop of apricots. When Margaret tried once to take an armload of firewood from her, Sadie rebelled, saying: "What are you trying to do, kill me? I need my exercise."

There are no other concessions to her age. She loves to dance. "I learned all the steps from my Uncle Charlie at the country dances in the basement of the Luna Theater in Clayton when I was a schoolgirl."

Mail is delivered three times a week and she reaches her mailbox at six miles. It's 83 miles to Clayton. She could take the shortcut, just 63 miles, but that's dirt road all the way and there's no cell-phone service anywhere.

At home in Long Canyon, she takes her evening walk. Deer, elk, and wild turkeys share the trails. Bears and rattlesnakes are more common than she'd like. A bobcat recently killed seven of eight kittens so now she and Margaret keep a new litter sheltered at night. Cats help keep the rattlesnake population down, but other wildlife keeps the cat population down.

Gloria and her husband, Jim, have always run cattle near Clayton. Asked about the future, Margaret says: "Both daughters will make sure this ranch keeps going. It's been in the family too long not to." And they'll have plenty of help. Sadie has six grandchildren, 16 great-grandchildren, and one great-great-granddaughter.

Last June five generations helped celebrate Sadie's 90th birthday at Freedom Outpost in Folsom. Friends came from near and far. Eloy Gonzales' band provided the music. The dancing went deep into the night where, more often than not, there was Sadie, in the middle of the dance floor, cuttin' the rug.

—Tim Keller

Nippers & Clippers

© Sheree Jensen

Sheepshearing at their grandparents' ranch in Pumpernickel Valley, Nevada, is one of Cody (age four) and Emily's (age two) favorite times of the year. Cody pushes the sheep up the shearing chute. After the wool is sheared he gathers it and carries it to the grading table. Little sister, Emily, jumps and plays in the soft pile of wool. Lots of cousins and family help out during shearing time. Cody and Emily are the children of Joe and Jamie Frey of Fallon, Nevada. Their grandparents are Gary and Billy Jo Snow.

Gwen Petersen, 86

You need a sense of humor.

A writer of ranch woman verse,
Some funny, some poignant, some terse
Intends to keep ranching
And laughing and dancing
Till taken away in a hearse.

"I get up every morning, look out the window, and wonder how I got so lucky," says Gwen Petersen of her home in the Yellowstone River valley near Big Timber, Mont.

Raised in Peoria, Ill., Gwen spent summers at her grandparents' farm, learning about farm animals and riding horses. From an early age, she determined to go West. Following her freshman year at college, she made a break for Montana. She earned a degree in occupational therapy at College of Puget Sound in Tacoma, Wash. She also lived in Arizona and Kansas, but Montana beckoned loudest. Gwen returned to work at the Montana State Hospital in Warm Springs. While there, she wrote skits and programs for the patients to perform.

Friends introduced Gwen to Martin Petersen, a quiet ranch foreman employed near Big Timber. As she tells the story: "Martin wasn't real verbal. I tried my best to carry on a conversation. He gave me a big kiss that night." At the time, Gwen was working approximately 150 miles from Big Timber. "Martin never went anywhere," she says, "but he came to see me."

The couple married and started a small outfit where the Boulder River empties into the Yellowstone. "We put together a piecemeal ranch, but that remained our home base," Gwen says. "My view of the world did a one-eighty. Tending cattle, horses, sheep, dogs, pigs, chickens, cats, milk cows, and planting and harvesting a garden fit me like a comfortable shoe."

Gwen continues: "No one could take care of a calf or a baby animal the way Martin could. He found lost sheep, pulled the babies from mama cows having a tough time, doc-tored them, made them live and thrive. And irrigate? Martin could irrigate a rock and make grass grow." He passed away in 1999.

Gwen has penned a plethora of humorous poems, columns, and books. In 1985, she was one of 11 women and a "feedlot full of guys" invited to perform at Elko, Nevada's first cowboy poetry gathering. She's been back seven times. Much to her surprise, she

"Martin wasn't real verbal. I tried my best to carry on a conversation. He gave me a big kiss that night."

was even asked to appear along with Waddie Mitchell on "The Tonight Show" with Johnny Carson. She recited her poem, "Hats for All Seasons." Don Rickles was also on the show that night.

Gwen has written a weekly column, "In a Sow's Ear," for 25 years and published more than a dozen books, including: "The Ranch Woman's Manual," "Everything I Know About Life I Learned From My Horse," and "How to Shovel Manure and Other Life Lessons for the Country Woman." Director and conductor James Stanard wrote music for four of Gwen's poems. As she says, "It's a thrill to hear one's poetry sung by a full chorale accompanied by an orchestra!"

Gwen is currently working on volume two of "How To Be Elderly, A User's Guide." She writes from experience. "I can't get on my saddle mare without a ladder," she says. "And if I were to get on, I wouldn't be able to get off." Both Gwen and the mare were more agile when they participated in Montana's

CLOCKWISE FROM LEFT: As Miss September in the 2006 "I See By Your Outfit" fund-raising calendar. ➤Martin and Gwen where they spent much of their time...in the corrals. ➤In February 2014, soon after celebrating her 86th birthday. ➤Gwen (standing in center) teaches student nurses about occupational therapy while working at the Montana State Hospital in Warm Springs. ➤Gwen bought a scooter and added a sidecar for Bailout the cow dog to take a road trip.

Centennial Cattle Drive in 1989.

What inspires Gwen today? "The world is a pretty funny place," she says. "It has its tragedies, but there's a lot of humor. And on a ranch or farm, you need a sense of humor to avoid snakes, skunks, and the occasional runaway tractor."

A road trip had been on Gwen's bucket list of things to do, so in 2011 she bought a scooter and sidecar for a central-Montana adventure with Bailout the cow dog. "A couple of friends were my support crew," she says. "We took a wrong turn and ended up seeing towns we hadn't planned on, including Grass Range."

An enthusiastic community supporter, Gwen says: "I love writing plays and skits and putting on shows. I have an endless stream of ideas. If I just had another life to live, I could get it all done. Maybe."

—*Jeri L. Dobrowski*

Bud & Dorothy Cashbaugh, 92 & 91

A rich tradition of ranching.

PHOTO COURTESY COOKE'S PHOTOGRAPHY, BISHOP, CALIFORNIA

After 92 years, James "Bud" Cashbaugh can still describe his childhood morning routine. "I'd help Dad with chores and then we'd take the bus into town because we were too far out of town to ride," he says. "But I remember the kids who rode to school, and there was a spot for them to tie their horses up outside the classroom."

While Bud rode along the dusty roads skirting Bishop, Calif., his future bride was playing in a cardboard box in a one-room schoolhouse north of town. Dorothy Amon was born in the bustling railroad town of Laws, where her mother, Mabel, was a schoolteacher and her father, West, worked at the Hughes' general store. Mabel rode her horse north of town every morning to the schoolhouse, instructing first through eighth grades.

PHOTOS COURTESY CASHBAUGH FAMILY

CLOCKWISE FROM TOP:
Granddaughter Cristina Giacomini Hughes' wedding on July 1, 2011, in Long Valley. Photo shows the family, except for son, James, and great-granddaughter, Kate. ►Dorothy and classmates at the one-room Riverside schoolhouse north of Bishop in 1931 where her mother taught school. Dorothy is riding in front on the light-colored burro. ►Bud and his sister, Imogene, at their home at Buckley Pond in 1926. ►Dorothy in 1931.

"Mom would bring me along and make a bed for me in a cardboard box," Dorothy says. "The older kids took turns watching me during the day."

Bud's family migrated to Owens Valley in 1865 and started to farm and raise beef cattle. They worked long and hard until the Depression. Families in the valley, along with Bud's, sold their land to the Department of Water and Power, City of Los Angeles, for water rights. The DWP knew water would become a precious resource for the growing urban population, and the agency found its answer in the lakes and rivers of the Owens Valley. Bud's family moved to town in 1930 and leased their land back, only running cattle because of the limited water supply.

Bud attended the University of Nevada, Reno, and was drafted and served in the Army in Alaska during World War II. He returned home to run the ranch with his father, his cousin Robert, and Uncle Gus.

Dorothy attended the University of California at Berkeley and Los Angeles, and then returned home. While Bud and Dorothy had known each other since childhood, fate would unfold one evening when they went on a first date at the Masonic dance hall. Bud wasn't keen on dancing, but found out quickly that he didn't have much choice in the matter.

> **"Mom would bring me along and make a bed for me in a cardboard box to sleep in. The older kids would take turns watching me during the day."**

"I didn't know how to dance. Well, she was just gonna force me to dance," Bud says. "And I thought, 'Well, you might force me, but it will be the last time!' I learned from that it's her way or the highway!"

While they may have gotten off to a rocky start, the couple married in 1950 and they were blessed with daughter, Alonna, in 1954 and son, James, in 1957.

Bud's father passed away in 1959 and in 1981 Robert and Gus retired, so Bud and Dorothy purchased their half of the ranch. Today they watch over the management done by their son, daughter, and son-in-law, Gary. The family winters their commercial cows around Bishop (elevation 4,150 feet) and summers them in Long Valley (elevation 7,130 feet) near Mammoth Lakes. Calves are sold

and shipped in October from Long Valley.

Bud and Dorothy also grow blue-ribbon vegetables, and Dorothy can often be found crawling on the floor with their three great-grandchildren, Will, Ben, and Kate. The rich tradition of ranching is steeped within the generations of their family, from their children and three grandchildren to the sixth generation of ranchers, those three great-grandchildren.

From the Depression to the DWP's purchase of the majority of land in Owens Valley, change has been a constant in their lives. While ranching has never been easy, they hold no regrets. Bud says, "If we had it to do over again, we'd do it the same way."

—*Cristina Giacomini Hughes*

You're Going the Wrong Way!

© *Cheryl Rogos*

Four-year-old Tel and his always-eating, trusty horse, Blue, are supposed to be riding point on the MLY Ranch in Greer, Arizona. Tel turns around to see that those calves are not following him like they should. The other cowboys were roping an angry cow that belonged to another ranch, dragging it to the trailer and all those calves had to have a look-see.

This was going on behind the scenes and these three cowboys will be back soon to help Tel. From left to right: Laden Summer, Todd Sweeney and KT Thompson.

Bert Smith, 93

Rancher, entrepreneur, patriot.

Bert Smith, age 93, still works six days a week, dividing his time between his sprawling military surplus and retail outlet near Ogden, Utah, and the Nevada ranch that he and his brother have owned since 1947.

"When I came home from the Marine Corps, I was bound and determined to get even with FDR for drafting me in the first place, an old man of 24 with a wife and a baby." When he was drafted, Bert says he had just 21 days to liquidate his trucking business and report to basic training. "It was just the first of many injustices by the federal government."

After the war, Bert went back over his old ranching area in Ruby Valley, Nev., selling things. His training as a cattle buyer and dealer, starting at age 17, prepared him for the day when an opportunity to buy a ranch presented itself. He is quick to credit the hand of Providence to explain how one of Nevada's substantial ranches came into his hands. He also credits his late wife for safeguarding their nest egg while he was away during the war.

"Amelia was the real hero," he says. "She saved every penny of the truck money and lived on $85 a month from my Marine Corps pay. The $10,000 was still in the bank and we put it all down on the ranch."

With a handshake and an agreement with the seller to pay the balance in cash in 30 days, Bert sold all the cattle, machinery, hay, and grain off the ranch, keeping only one tractor. "The big problem then was to get the livestock back to make it a cattle ranch. I continued buying and selling government surplus to get my capital back to stock the ranch, and as a result I developed two businesses at once—the ranch and the military-surplus business in Ogden, Utah, named Smith and Edwards."

Bert's brother, Paul, is a full partner in the Smith Brothers OX Ranch. "It took us the first 10 years of ranching to get a herd built back up," Bert says. "While I was running the store and busy in the freedom business, Paul was running the ranch. I couldn't have asked for a more loyal brother."

When the ranch came into the Smith brothers' hands, it had been sold three times in three years and the grazing permits had been reduced 10 percent each time. When Bert objected, the Forest Service told him it had the power of eminent domain to take any public land it wanted.

"I began a search to find out why and how the ownership of state lands had gone wrong and turned to the U.S. Constitution. That changed

With 21 days to liquidate his trucking business and report to basic training, Bert says, "It was just the first of many injustices by the federal government."

the course of my life. I found the answer in Article 1.8.17, which said the federal government only had jurisdiction in 10 miles square and was not allowed to own land inside of a state unless it had permission and paid for it."

This discovery led Bert to a lifelong study of constitutional principles that continues to this day. He served as chairman of the Public Lands Committee for the Nevada Cattlemen's Association, writing a resolution that became known as the Sagebrush Rebellion. After nearly 50 years fighting for state lands, Bert is optimistic that current efforts to transfer public lands back to western states will ultimately be successful. "Organizations

PHOTOS COURTESY BERT SMITH

CLOCKWISE FROM TOP: Bert with his dog, Pup. ➤ Bert and Amelia Smith with children in 1964. From left: Steven, Brent, Sharlene, Sherma, Ruth Ann and Jim. Amelia died in 2004. ➤ Bert and Kathy celebrate their sixth anniversary in 2012.

such as the American Lands Council are making headway in educating governors, legislators, and especially county commissioners. I've traveled many miles with many good friends over the years teaching county governments how to pass constitutional law into their county ordinances."

Bert is still buying and selling surplus and cattle 66 years later. His secret to making money? "Go to work!" His simple Economics 101? "Buy for one, sell for two, never pay interest, and get right with the Lord. He'll take care of you."—*Kathy Smith*

Edge of the Black Rock

Homeschooling in the middle of nowhere.
Words and photos by Katie DeLong.

It is said that little boys are made of "snips, snails, and puppy-dog tails." I don't believe whoever wrote that meant it literally, but that's just because they had never met one particular little boy—one who really does have a puppy-dog tail hanging on his bedroom wall.

No, this was not the misfortune of some

Homeschooling for us was more of a necessity than anything else. Although a small, rural, one-teacher school exists in our area, it is 65 miles and one very long bus ride from our cattle ranch in northern Nevada. The nearest town with a grocery store is 90 miles away. Basically, we live in the middle of nowhere, right on the edge of

go to school, because I already know everything." Oh boy, this was going to be a long year.

We started homeschool a week earlier than the regular public school. I thought that would be best since we'd be missing a week of school time to show our horses at the county fair in September. What's more, there would be weaning, cow work, and a few fall brandings to schedule around.

We began the first and every day of kindergarten with the Pledge of Allegiance and a prayer. (The Lord knows we were going to need a lot of them.) Then we discussed homeschool rules: listen, be nice, work hard. (Interpretation: do as Mom says, or else!) Reading, 'riting, and 'rithmetic were daily necessities, while science, art and the like were built in as we went along. I was amazed at just how much material we could cover in a day. That was as long as my easily distracted son didn't goof around too much, or we didn't have too many outside disruptions!

Throughout the fall, school seemed to be progressing along well. The baby owlets that had tumbled out of the haystack prompted a great science unit on barn owls and other predatory birds. The frogs and toads the kids caught in the fields and creek provided a great science discussion on amphibians. My husband, Will, was also a good sport about providing "learning opportunities" for the kids—everything from bug collecting to overseeing the papier-mâché volcano and erupting it with vinegar and baking soda.

Then, of course, there was the butchering. Whenever we butchered a beef, the kids got hands-on experience with anatomy. You can rest assured that all ranch kids know exactly where their food comes from.

My middle daughter, Matti, though not yet in school, was learning right alongside her brother. The baby, Louise—whom her brother had insisted be named Buck Deer DeLong if she'd been born a boy—was also along for the ride. We learned how to be flexible and would take recess to change the baby's diaper, chase the horses out of the yard, or help on

The ranch is 90 miles from a grocery store, located on the western slope of the Jackson Mountains. The peak on the horizon is Big Mountain, part of the Black Rock Range.

poor, mutilated creature, but rather the result of a mishap from a curious, unsuspecting canine that happened to be in the way of a charging bovine. The steer stepped on the puppy's tail, severing the end, resulting in a yelping but none-the-worse yellow Lab that had learned a valuable lesson that day in staying out of the way. The young boy, although worried about his precious dog, immediately picked his newfound prize out of the dust. With the help of his grandfather, he tied a string around the tail and hung it in a place of honor on his bedroom wall.

If I had ever thought that I'd be the mother of such a fun but outrageous little boy, I would have never believed it. Let alone would I have believed that one day I'd be homeschooling such a child—that child along with his two blond-haired, mischievous little sisters.

the Black Rock Desert.

Having been a public-school teacher, I was intrigued yet still nervous at the prospect of teaching my own children. As a classroom teacher, I was in a position of authority. When I spoke, children sat up and listened. Didn't they? How come my own children never seemed to listen?

If my kids wouldn't listen to me, how was I going to teach them? Then there was the issue of stereotypes and stigmas attached to homeschooling. Were my children going to end up labeled as those "weird homeschool kids?" What about socialization? We love all our animals on the ranch, but don't kids need other children to play with? As these questions and uncertainties swam in my head, my son, Billy, didn't help the situation when he said: "I'm not like other kids. I don't need to

As these questions and uncertainties swam in my head, my son, Billy, didn't help the situation when he said: "I'm not like other kids. I don't need to go to school, because I already know everything." Oh boy, this was going to be a long year.

Billy rides for pleasure and to help his father on the ranch. RIGHT: Matti feeds "favorite horse" Smoky. BELOW: Louise likes hats, just like her great-grandma Rosita.

the ranch when needed. We'd take a break for lunch, too. One day I asked Matti if she'd like a grilled cheese sandwich. She said: "Yes, I don't want a *boy* cheese sandwich. I want a *girl* cheese sandwich."

As winter approached, we continued our daily lessons as well as our one-day-a-week visit to "town school." Early in the school year, it became apparent that our son would need a little extra help with speech. After a public-school evaluation, I began driving Billy to speech once a week in town, 90 miles away. Afterward, he was permitted to visit kindergarten. During this time, I could pick up groceries and complete other errands.

The arrangement was working out nicely and we were very pleased, not only with our son's progress in articulation, but also with

socialization. The benefits of this really became apparent when one day Billy began to tell me about one little boy who refused to stand by him in class or in line on the playground. Billy explained that every time he stood next to him, the boy would run away. I asked Billy how this made him feel.

"Sad," he said.

"Why do you think this boy won't stand by you?"

"I don't know. He's mean, I guess."

"Well, what do you do when he runs away from you?"

"Oh, I follow him."

Of course this led to an in-depth discussion on *not* following people if they do not want to be around you, instead finding someone else to stand by or play with. Although, like all mothers, my heart breaks to see my children hurt or sad, some of life's lessons just have to be learned the hard way. Lessons learned here: first, children can be mean; and second, don't annoy them.

In addition to homeschool and one-day-a-week town school, we were invited to participate in special activities at the rural school in our area, including basketball, special classes, and the Christmas and spring plays. These were all fun learning experiences.

As Christmas came and went, we were relieved to get our Christmas break—especially the homeschool teacher. Although I was enjoying teaching again, there were definitely days that I thought I might pull out all my hair. Teaching your own children can be very frustrating at times!

In spite of this, more and more I found myself feeling proud of the progress we were making, both in academics and in balancing everything else—ranch work, social activities for the kids, cooking, housework, etc. Although my house often got much dirtier than it ever had been before, some things just have to go by the wayside. Moreover, the baby was running and climbing by now, so nothing, including schoolwork, was safe from complete and total destruction. Cleaning up messes seemed to be a full-time job.

(Most mothers know this already.)

Springtime brought branding, riding, and gathering, which meant homeschool had to be adjusted to make it all fit. Our spring science unit on plants culminated with the planting of our own garden, and Billy's daily writings often focused on his ranch adventures. We seemed to make it all work, though, and even got a good jump on first grade.

On May 31, 2012, at a ceremony in town, Billy graduated from kindergarten with

grandparents and other family members in attendance. We had done it! We had made it through our first year of homeschool. Who would have thought we'd come so far? (Of course, this *was* only kindergarten!)

As our pride swells, the looming question remains: Will we continue to homeschool? I suppose only time will tell, but for now, as long as it is working for us and for our situation, we will keep plodding on. Plus, we've got lots of treasures and memories from our first year of homeschooling to show for it, including that puppy-dog tail and a kindergarten diploma—both displayed in a place of prominence on the bedroom wall of one particular little boy. ∎

Katie Marvel DeLong is granddaughter to Tom and Rosita Marvel (see page 34).

Victor Garber, 95

Spuds, turkeys, sows and
sheep: a great way to live.

Turkey ranching in Wyoming? Victor Garber remembers it well. "I grew up on a typical ranch in the 1920s where we had chickens, calves, bum lambs, pigs and turkeys.

When I was in grade school in the 1920s and early '30s, I herded between 500 and 1,000 turkeys outside of Sheridan. I'd herd them into the fields where they'd eat Mormon crickets and grasshoppers and roost in the trees. I had a Shetland pony and a couple of dogs for help." They went to a turkey cooperative where birds were slaughtered and rough dressed a few weeks before Thanksgiving and Christmas and shipped in boxcars to the eastern markets. "They weren't even refrigerated."

Victor was born on Sept. 14, 1919, in the log cabin in the yard where his son, Roy, currently lives. The 95-year-old remembers riding to school horseback because there were no school buses. "The school had two horse barns," he says, "which were sheds with stalls. They were good, tight buildings so the animals were well protected from the rain and snow."

While World War II was raging, Victor and his brothers were busy raising food. They ran sheep, cattle, and hogs and grew sugar beets, potatoes and navy beans for the war effort. "For three years of the war, Big Horn High School was dismissed during harvest so the kids could help with the sugar beets and potatoes."

Sheep were abundant in Sheridan in the 1940s through the 1970s, and Victor served on the predatory animal board of Sheridan County for 43 years. It existed to control coyotes. "Although predator control is still alive and well, it was especially critical when sheep were the major livestock in Sheridan. The coyotes aren't quite so bad now that everyone raises cattle."

Most ranchers had Hereford cattle in

CLOCKWISE FROM ABOVE: Victor and Patty have been married for 10 years. ➤ Victor and his late wife, Phyllis. ➤ Victor served as a Wyoming legislator, first as a Democrat and later as a Republican. ➤ Jim Roush and Victor (right) stand beside a rock crusher. ➤ Orr stands behind brother Victor, ca. 1922.

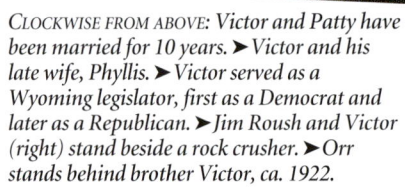

"We ran 100 brood sows in herds on irrigated alfalfa, and we never fed those sows grain during the summer. We got to be real fencing pros. Our hog roundup was wild!"

those days. There were quite a few Holsteins, as Goose Valley had a lot of dairies. "By the end of the war years, my brother, Orr, and I were running 2,000 head of sheep, 250 head of cattle, and 100 brood sows, along with growing potatoes and feed," he reminisces.

"We never ran hogs in confinement. We ran them in herds on irrigated alfalfa, and we never fed those sows grain during the summer. We got to be real fencing pros." He laughs. "Our hog roundup was wild!" They phased out of the hog business about the time they started their feedlot.

"My brother and I and our wives bought our current ranch in 1942 and took possession of it in 1943. My wife, Phyllis, myself, and baby Paul moved onto this place the day Franklin D. Roosevelt died."

Victor and Phyllis incorporated Garber Agri-Business when they real-

ized their three sons were going to be part of the family outfit. Today, David runs the cow-calf operation. Roy ran the feedlot until it closed a few years ago but is now involved in all facets of the business. Youngest son, Paul, runs the construction business.

The venerable rancher served in the Wyoming Legislature as a Democrat in the House of Representatives in the early 1960s, and then he ran and won as a Republican in the 1980s. Victor actively helped on the ranch until a year ago when he contracted pneumonia.

The family isn't afraid to change with the times. "You need to keep up with the trends. If you don't move with the ag economy, you'll go broke," says Victor. He can't say enough about the benefits of ranching. "It's a great way to live and the best way in the world to raise children. There is only one thing you can give children, and that is responsibility. Working in agriculture teaches them that."

—*Rebecca Colnar*

Troy McNaught Westby, 96

Skunked at school.

Troy McNaught Westby was born Jan. 5, 1916, on a farm east of Glasco, Kansas. "In 1933, I married and moved to the Nebraska Sandhills. In 1934, our only child, William 'Slim', was born and the following year we moved to a ranch southwest of Wanblee in South Dakota on the Pine Ridge Indian Reservation. Our log house sat on the bank of a creek, with the barn and corrals on the opposite side. A large cottonwood tree laid across the creek for a footbridge."

Troy had three years of high school "normal" training in Kansas, which in those days qualified her for a teaching permit. In 1941, the now defunct South Dakota county of Washabaugh was hard up, with no money to pay teachers, so she agreed to teach at Eagle Nest School for $90 per month with a permit and agreement to get her high school diploma, which she did while teaching that first year.

That one-room country school was eight miles northwest of the ranch and Troy rode horseback to it on Monday morning and back home on Friday evening. She taught there for several years, staying in the schoolhouse behind sheets hung on wires or occasionally boarding with a nearby family.

"One winter day in the mid-1950s, two little first-grade boys were the only students to get to school. Snow was piled deep in the road and on most of the school ground. After lunch, they went out to play in the snow. I was checking papers at my desk and keeping an eye on them through the windows."

Suddenly, they started running and floundering through the snow toward the building and she ran to the door. "I could see a skunk behind the boys. Whether it was chasing or following them I didn't know, but I knew it would come into the building behind the boys."

She looked around for a weapon and found a baseball bat. She thought, "I only have one chance…" and swung the bat and hit the skunk on the head. "It dropped into the snow, kicked a bit, then a yellow substance began to spray out on the snow. The odor was horrible! The boys stopped crying and we all three stared in stunned amazement."

Troy says they didn't take many pictures in the old days and most were taken by visitors or relatives. "We figured out that it was probably around 1956 before we had a camera in the family." CLOCKWISE FROM LEFT: Troy rides Shorty in 1941. "He was coal black and in the fall his hair got curly, then slicked out in spring. He liked pushing baby calves along in the drag when we trailed to summer pasture. I see that saddle blanket is slid back, and it is kinda worn lookin'." ➤ Four generations in 1916. Six-month-old Troy sits in her mother's lap, with grandmother and great-grandmother. ➤ Sisters in 1946. From left: Bert, Troy and Gert. ➤ Troy in 2011. She says, "Seems like I can remember the horses way better than some of the people back then."

> ## "I could see a skunk behind the boys. Whether it was chasing or following them, I didn't know."

The boys didn't want to get near the animal so the courageous teacher dragged the carcass off the school grounds, then buried it in snow. "In those days," she says, "you never knew what to expect when teaching in a rural school."

The family moved into a frame house in Buzzard Basin where they lived until they moved to town in 1956. Four years later, Troy started teaching in the Rapid City school system and continued learning herself. When she retired in 1978, she held a composite major in education and a major in art. Slim married and took over the ranch in 1953. He sold it in 1976 and started making saddles.

Troy now lives in New Underwood, where she writes poetry and short stories and frequently entertains at area functions. "I have been a prolific writer all my life, with shelves and boxes of writings, plus many that have been lost or destroyed over the years."

She admits it wasn't always easy keeping material safe from little varmints in the old log houses of the time, plus the weather that came in through the cracks when the chinking crumbled and fell out.

"I spent a lot of time mixing mud from the buffalo wallows and chinking the openings between logs in that old ranch home."

—*Yvonne Hollenbeck*

FALL 2019

The Horseman's Protégé

With 70 years separating them, this cowboy and his young friend are drawn together by an inspired flair for horses and mules and an abiding respect for one another. By Marjorie Haun

Ked Somerville

Kedric "Ked" Somerville's ranching roots go way back in southern Utah history. He was born in 1936 and raised on his large family ranch that covered some of the prettiest acreage in San Juan County. His dad was a partner in the Scorup-Somerville Cattle Company and when Ked was a young man they ran cows on the Spring Creek part of the operation. Traversing the slopes of mountain ranges, slickrock arroyos, and vast stretches of sagebrush and pinion country, the S&S Cattle Company ran up to 10,000 head at a time.

While a tender adolescent, Ked struck up a friendship with Zina "Marleen" Rasmussen, a girl also born in 1936. Her family were wheat farmers and owned a place near Peter's Point, north of Monticello. Although his home was some 10 miles away, Ked would look out towards the Rasmussen farm and think of that sweet Marleen. Whenever he could find a suitable excuse, he rode the 10 miles to drop in on her and her family.

"There were so many deer in the hay fields," Ked says, "so one time I shot a doe and gutted her out right there in the field and sacked up the carcass and tied her on behind my saddle." He continues with a smile, "Then I took that deer to Marleen's dad, kind of like a token of trade...a buckskin for your daughter."

At 21 he married Marleen, and they both attended Brigham Young University for a year.

> **As providence would have it, before Marleen left his side, Ked met Boden Carpenter, who was just five at the time. Boden would come, not to fill a void, but to create a new space in Ked's heart, and bring a different kind of love into his life.**

They decided then to start a ranch of their own. Ked was serious-minded and determined to put out the best cattle he could.

Ked and Marleen raised a girl, Tracy, born in 1957, and a boy, Kody, born in 1959. For 40 years, Ked ranched and raised his own line of registered Simmental cattle. Although never more than 50 to 60 head, his little herd had premium DNA. Ked's reputation was such that he had cattlemen purchasing bulls from his ads in the Utah Farm Bureau newspaper, sight unseen. While he was breeding and selling his bulls, Marleen watched over the farm operation. Ked says: "We were irrigating and growing hay, and Marleen was right in the middle of all that. When I needed help she was there."

In April of 2011 Marleen was diagnosed with lung cancer. "The doctor gave her all the alternatives for treatment," Ked says, "but she just didn't want to go through all that." After five months, Marleen passed away on Sept. 1, 2011.

With the loss of Marleen, Ked sold his cows to a family in Spanish Fork which had been building its herd from his own Simmental stock. He describes the transaction: "They came down to get some heifers and when they got here they looked over the fence and said, 'How about your herd?' and I said, 'Okay, let's make a deal over this gate right here, right now,' 'fore I change my mind.'"

It was a provident moment for Ked. "These folks loved my cows and, in fact, now they have one of the best Simmental bull sales in the state." And as providence would have it, before Marleen left his side, Ked met Boden Carpenter, who was just five at the time. Boden would come, not to fill a void, but to create a new space in Ked's heart, and bring a different kind of love into his life.

Bode Carpenter

Boden "Bode" Carpenter was born in 2006 and as a newborn was adopted in Detroit, Mich., and taken by his new parents to their home in Kearns, Utah. The story goes that his parents thought they would be adopting a girl, but what they got was a boy with unbounded giddy-up and enormous nerve.

Bode's family was not content in Kearns so they moved south to Monticello to be close to his grandparents. "One day," Bode says, "I was sitting watching TV and I was bored, so my grandpa said, 'C'mon, I'm going to take you down to Ked's.'" Bode asked, "Who's Ked?" and Grandpa said, "Never mind, just get in the truck." That day, little Bode's life changed forever.

Ked happened to be an old friend of Bode's granddad. Though he had sold his cows, Ked kept his horses and a couple of Tennessee Walker-cross mules, a molly named Kick and a 17-hand john named Honk. "When we got to Ked's I saw Honk and I petted his nose and that was it," Bode says. "I just never left. I kept coming back and coming back."

A Rare Friendship

Bode's important training has been at Ked's hand, and his affinity for the corrals and the horses has resulted in a lot of derring-do. When he was seven, Bode rode on Kick along with Ked up to Peter's Point. "I didn't know what I was doing. I had no clue, and there was this big ledge the mules had to jump up onto, and the mule took a big old hop and it scared me so much I just wanted to get off and walk home."

Ked adds: "I had no idea what was going through his mind. I just looked back and he was still in the saddle so I kept on going."

For his baptism at eight years

of age, Bode was given a Shetland-quarter horse mare named Ruby, and he's now in the process of breaking his own palomino colt.

Bode's hardest lessons, however, have been on the range, cowboying for the J-Bar Ranch near Mexican Hat. At 83, Ked still hires out as a cowboy, and Bode is his sidekick. According to Ked, together they put on between 1,000 and 2,000 miles a year pushing cows and trail riding, primarily on the backs of Honk and Kick. With unwavering

trust Bode is always at Ked's side, learning not by verbal instruction, but by careful, daily observation. Growing tall and strong and with hands bigger than most men's, Bode never shies away from challenges, and has become adept at roping, even while standing on Kick's saddle.

Ked smiles and says: "He's just an acrobat. He ain't broke anything yet."

Bode is a gifted horseman. But more important than his passion for horses is the quality of his heart. Ked says: "He's not like any other boy I know. He's so courteous. When we go to town he's always looking for a lady so he can open the door for her. Or if we're at the post office and someone drops something, he'll pick it up for them. It's just built into him, you know. I've learned a lot about how to be a decent person from Bode."

Ked and Bode, though generations apart, are cut from the same resilient and colorful cloth. To Ked, his time spent with Bode is an investment in the future of a fine young man. To Bode, Ked is a fountain of knowledge, a patient mentor, and a jovial partner. With a rare friendship that transcends age and the mundane limits of life, Ked and Bode exemplify the adage, "The best thing for the inside of a man is the outside of a horse." ∎

Marjorie Haun lives in La Sal, Utah.

CLOCKWISE FROM TOP: Literally learning the ropes at age seven, Bode urges Kick through a mud hole on the range. ➤ Each year Ked and Bode hire out to the J-Bar Ranch to move cows through the Utah desert. ➤ On his Monticello farm, Ked talks horses and life with Bode. ➤ Bode practices roping on Ked's home-crafted steer.

The Strength of Family

© Larry Angier

Waylon S. Dell'Orto, age three, keeps a good hold on his tiny brother, Knox, at his first calf branding in 2017. Waylon S. has been at brandings for several years already and wants to show Knox the ropes.

These brothers are sixth-generation ranchers, whose families have been grazing cattle in California's Sierra Nevada since the 1800s, from the foothills of the Mother Lode to the High Sierra basins in summer. Nine sets of grandparents—back to the great-great-greats—have neighbored, driven cows, branded and celebrated their lives together in the best of ranching tradition.

Bob Hanson, 96

`Starting early.`

Bob Hanson is a son of the land, a war hero, and a product of the tenacious folks who settled the country when work was done by hand, on horseback or with a team.

One of seven children born to Jim and Elsie Hanson, Bob grew up on the family homestead near Bison, S.D. His father freighted with a four-up team of Percheron horses, and also had a custom harvesting outfit with a header pulled by six horses, and two header boxes, also pulled by six horses apiece.

"At eight or nine years of age, I was put to work. I halter-broke all the colts to stand tied at the hayrack all day while the mares worked. Those colts learned to stand patiently until we broke at noon to eat, then they got to nurse and be with the mare until she went back to work."

His father stood three stallions back then, a thoroughbred, Percheron and Shetland. "He bred the thoroughbred to the Shetland mares and I broke them as a kid. My oldest brother, Bud, would ride them until they quit bucking, then I took them and made kid ponies out of them."

After country school, Bob attended the state agriculture school and learned blacksmithing, welding, carpentry and other trades. He eventually worked as a hunting guide, brand inspector, deputy sheriff, orchard worker, and sheepshearer.

World War II called and Bob, 23, answered in 1941. He was placed in the 15th Cavalry Reconnaissance Squadron at Fort Riley, Kansas, where he rode and trained horses every day. Ranch work and hunting paid off when he scored third out of 300. He went to three gunsmithing schools while in the Army. He also had 87 contested fights in the boxing ring.

In 1944, in Doslet, Brittany, France, a 40mm incendiary shell hit the armored car he was riding in. The driver was killed and the other men and Bob were burned, plus Bob was wounded by shrapnel. "It busted up my feet so bad I didn't think I could even walk, but when the artillery shells started going off, I ran like a squirrel," he says. "I couldn't get away though and the Germans

PHOTOS COURTESY BOB HANSON

© JAN SWAN WOOD

> **"I didn't think I could even walk, but when the artillery shells started going off, I ran like a squirrel."**

caught me. A German aid man who was very good with burns went to work on me and probably saved my life."

A nearby house had been converted into a hospital. "They hauled me in a wheelbarrow 'cause I was hurt and burned so bad I couldn't stand to be carried. A French doctor worked on me too and said I had a one-in-a-hundred chance of survival."

After treatment Bob was sent to a German prison camp. "My colonel was captured just before I was and he spoke fluent German so could deal with our captors and mess with their heads and protect us a lit-

tle." He marched for miles on his crutches with his wounded feet. "You think you can't walk any further, but then they'd pull the scabbard off the bayonet on their rifle and press the point to your back and you find that you can keep going after all." When freedom finally came, Bob weighed just 87 pounds and his feet were crippled.

Upon his discharge in 1945, Bob went to gunsmithing school in Denver, Colo., and

CLOCKWISE FROM TOP: Bob was in the 15th Cavalry Reconnaissance Squadron at Fort Riley, Kansas, and is one of the troops in this 1942 photo. ►Bob and Donna on the day they were wed. ►The last time Bob rode he was 93. His last good horse, a stud named Cordy, died in 2000. ►Bob Hanson, 96, is a decorated veteran, a cowboy and craftsman, and a fascinating storyteller. He builds branding pots for area ranchers using his own design. The pots can keep a multitude of irons hot at the same time.

later married Donna Wright. They had three sons, Terry, Jim and Rick, who have all passed away. Donna died in 2002. He has 12 grandchildren, "a passel of great-grandkids, and one great-great-granddaughter."

Bob stays busy now building branding pots and artistic pieces in his blacksmith shop and maintains a unique collection of cowboy items, including horsehair ropes, bridles, bits, bosals, different types of catch ropes and other memorabilia. Bob rode until he was 93 years old. "I lost my last horse, a stud named Cordy, four years ago."—*Jan Swan Wood*

Andy Little, Sheep King U.S.A.

Two dollars, two dogs and a long way from home.
By Teri Torell Murrison

Once upon a time, a strapping young man dreaming of a better future left his Scottish homeland for the high desert rangeland that is southern Idaho. His story is the stuff of screenplays, or should be. In 1894, with his father's blessing, Andy Little bought a trans-Atlantic ticket to America. Along with his herding dogs, Jim and Katie, he was armed with a letter of introduction to someone who knew someone and $25 from selling a band of sheep.

He was one of a number of successful immigrants to the West who started out as sheepherders before becoming ranchers and businessmen. He started on The Aikman, a ranch northeast of Emmett, Idaho, owned by fellow Scot, Bob Aikman. Once there, he watched for opportunities and took most of them.

Over the last century, Andy and his descendants have had a part in many important decisions about Idaho's future. Andy's grandson, Harry Bettis, 83, is still very much in the room. A successful rancher, bank director, philanthropist, and ardent defender of western heritage, Harry remembers Andy fondly, if realistically. Andy was tough, a hard man in business, with a soft spot for his grandson.

"I was lucky," Harry says. "Granddad was plumb good to me. I was the first grandchild. When I was about five, there was some emergency at the office and my Aunt Jessie called

to speak to him. Grandmother told her, 'Well, he's playing hide and seek with Harry now.'"

When young Harry would ride along with Andy and his driver, Harry took great pleasure in locking the two out of the car. Andy always relented and paid him 10 cents to buy their way back in, half of which bought Harry an ice cream cone and the rest a cigar for his granddad.

When Andy arrived in Caldwell in 1894

on the train, he had just two dollars to his name. It was not enough to ride the stage, so he walked the 20 miles to The Aikman in his wool suit. It was the kind of desert day where heat shimmers above the ground. The slope of the broad valley was gentle, but the foothills were steep. Few trees grew between Caldwell and the ranch. So he followed Willow Creek up.

When he came upon a man and dog try-

Andy did well in America. This is a family portrait taken in 1913. From left: Robert, Andy, Drew, Jessie, Adis, Agnes, and David. AT TOP: *One of Andy's herders, thought to be a relative, stands with Little dogs and sheep. It is a long way from Scotland.*

ing unsuccessfully to drive sheep across Willow Creek, he waded in to help. The man saw how well Andy's dogs worked and wanted to buy them. Andy sold him Katie, but he and Jim continued on their journey. He needed at least one good dog. Sentimentally, he went on to own a number of good dogs named Katie and Jim.

Andy's first job was to herd sheep, 1,200 ewes to a band in those days. Most had single lambs. He lived in camp with the herders and learned a lot, managing his grazing to make sure the sheep didn't run out of green grass. Over time, he learned to breed selectively, good ewes to good rams, and improved their feed to increase the incidence of twins.

Herders were typically paid once a year. When Bob offered him either a first paycheck or half a band of sheep, Andy was at a crossroads. Aikman offered to sell him the other half band on a note if he could get a loan from the bank. He chose the half band and the note.

From his start on The Aikman, he bought more sheep and began buying land. At one time, he owned 27 ranches and grazed 100,000 head of sheep on deeded and public lands. In 1929, he produced a million pounds of wool. He had team after team of beautiful Belgian horses and at least a thousand sheepdogs. Andy employed large crews of loyal Basque, Scottish, and Tennessean sheepherders, many

of whom remained with him for years.

His greatest tool was the herding dog. A good one could take the place of several men, protect the herd from predators, and enable his herders to run more sheep. He insisted that his herding dogs were not pets. However, a few rode everywhere with him in the car.

> **"They said that he could go to five sheep camps and get the list of groceries from each one, never write anything down, and remember it all. He had a brain like no one else."**

With up to 400 employees at one time, he developed a long list of rules to govern life and livestock husbandry on his ranches, from what time to get out of bed, to when to feed horses, dogs, and sheep. He provided typically unbending direction for his employees to follow. Nothing escaped him.

"He had the damnedest vision and attention to detail you ever saw," Harry says. "He could see a string of horse teams one day and if there was a different team swapped in the

next day, he noticed. He could go to maybe five sheep camps and get the list of groceries from each one, never write anything down, and remember it all. He had a brain like no one else."

In 1901, Andy's father died. While he was in Scotland to see his mother, Andy visited a family that lived 50 miles from his hometown. Agnes McMillan Sproat—Adis for short—was out in a hayfield working. She caught his eye with her strength, her skill with the hay rake, and with her looks. They were both smitten. They married in New York in 1903 and came back to Idaho.

Adis kept a large garden, chickens, and a few milk cows. She was known for her intelligence, industriousness, and for entertaining. Over the years, the Littles hosted senators, governors, and sheepherders in their home. Both Adis and Andy were prolific knitters (and highly competitive with each other). Adis enjoyed playing sports and fishing and was considered a pillar in the Emmett Presbyterian Church.

Eventually they moved off The Aikman to Emmett, where they raised five children: Agnes, Jessie, Andrew (Drew), Robert, and David. Adis and Andy's first home in Emmett

Responsible for his great fortune, it's only fitting that sheep grazed the front lawn of Andy Little's mansion —called the Forever Home—in Emmett. RIGHT: Andy and his oldest daughter, Agnes. Educated at Wellesley, she came home to work in the ranch office before marrying and moving to a ranch in the Wood River Valley. BELOW: In his management and his penmanship, Andy was precise.

was a large farmhouse, half of which was later moved to The Aikman, where it remains today. Jessie told of eagerly waiting in the street with her sister for Andy to return on his horse and give them a ride home, walking beside them. Andy was away building an empire much of the time, but he had a soft spot for his children.

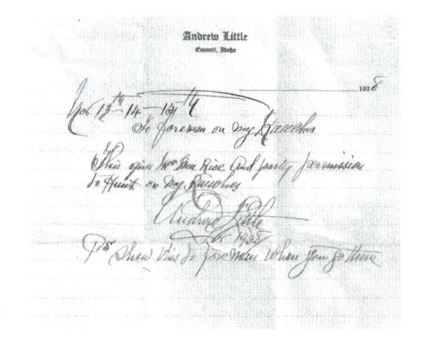

Agnes, the eldest, was Harry's mother. She married Laurence "Docky" Moore Bettis, a rancher, banker, and the son of Boise's first dentist. The couple had two children: one who died at birth and Harry. They moved to a ranch in the Wood River Valley where Harry was raised. Harry would marry Carol MacGregor and have three daughters: Laura, Catherine, and Janelle.

In 1924, Adis and Andy moved into a grand place in Emmett that Andy had built out of concrete. They called it the Forever Home. It's said that Andy, in response to the San Francisco earthquake and constant threat of range fires in the area, wanted a house that wouldn't fall or burn. The walls were rumored to be filled with old farm equipment for strength.

Harry's prize possession is a set of horsehair reins made by a well-known horse thief while in prison. When the man got out, Andy hired him so he wouldn't steal any of his horses. "One day Granddad asked me what I wanted for Christmas and I said I wanted those reins." He still has them.

By the early 1940s, Andy's health began to fail. He suffered a heart attack and the family convinced him to spend the winter in Santa Barbara with Agnes and Docky. He died in Santa Barbara in 1941. Adis lived another nine years before succumbing to complications after gall-bladder surgery.

Andy read the Congressional Record every day. "He watched things. He knew when land was going on sale." And he acted on what he knew.

Andy's portrait hung in the Saddle and Sirloin Club at the Chicago stockyards and is now in the Agricultural Hall of Fame in Kentucky. "I don't know if there was anything he set out to do that he didn't accomplish," Harry says. "He was most proud of coming here broke and winding up with 100,000 sheep."

In the late '80s, The Aikman became an Idaho Century Ranch, in operation for 100 years or more. The family threw a party and invited everyone who had ever worked for Andy. Over 500 people came. ∎

Teri Torell Murrison, a sheepman's daughter herself, writes from her adopted home in Eagle, Idaho.

Harry Bettis, decades later, treasures his granddad's horsehair reins. He still raises a few chickens on The Aikman and, following his grandfather Andy's tradition, all of Harry's ranch buildings and corrals are painted red.

The Seven Stages of Aging on Horseback

Words by Cindy Hale. Illustration by John Bardwell.

Stage 1: Fall off pony. Bounce. Laugh. Climb back on. Repeat.

Stage 2: Fall off horse. Run after horse, cussing. Climb back on by shimmying up horse's neck. Ride until sundown.

Stage 3: Fall off horse. Use sleeve of shirt to stanch bleeding. Have friend help you get back on horse. Take two Advil and apply ice packs when you get home. Ride next day.

Stage 4: Fall off horse. Refuse advice to call ambulance; drive self to urgent care clinic. Entertain nursing staff with tales of previous daredevil stunts on horseback. Back to riding before cast comes off.

Stage 5: Fall off horse. Temporarily forget name of horse and name of husband. Flirt shamelessly with paramedics when they arrive. Spend week in hospital while titanium pins are screwed in place. Start riding again before doctor gives official okay.

Stage 6: Fall off horse. Fail to see any humor when hunky paramedic says, "You again?" Gain firsthand knowledge of advances in medical technology thanks to stint in ICU. Convince self that permanent limp isn't that noticeable. Promise husband you'll give up riding. One week later purchase older, slower, shorter horse.

Stage 7: Slip off horse. Relieved when artificial joints and implanted medical devices seem unaffected. Tell husband that scrapes and bruises are due to gardening accident. Pretend you don't see husband roll his eyes and mutter as he walks away. Give apple to horse. ■

Cindy Hale lives in Norco, California. She wrote this after she and a friend lamented about how their bodies don't withstand the insult of getting tossed from the saddle anymore. This ran in Horse Illustrated in January 2011.

Norma Peters, 93

Crazy about horses.

The first thing you notice when you walk into Norma Peters' house are the photos, trophies and ribbons. The photos, mostly black and white, show a young girl with various stages of horse flirtations. In one, three-year-old Norma is alongside a workhorse. In many of the others a pretty young woman, wearing a white shirt and narrow tie, sits poised in the saddle as her horse soars over a five-foot gate.

"I was just crazy about horses," Norma says, moving about the room with a level of energy and enthusiasm not common for a 93-year-old.

Minutes later, seemingly oblivious to the evening chill, she's out the back door ambling to the barn and hollering, "Sox!" In a jiffy, the 25-year-old quarter horse gelding turns away from his feed and trots out to a beaming Norma. "He follows me around," she says, obviously pleased with herself and Sox. "He'll leave his hay and come when I call."

Sox is the latest in Norma's stable of well-loved and equally well-trained horses. When she enters the corral carrying two sugar cubes, Sox sidles alongside, nudging and cuddling her while she banters and giggles.

"I had to live someplace with my horses," she says to explain why she moved from central California to Fort Bidwell 10 years ago. The tiny town is a historic but sparsely populated community tucked into far northeastern California and just a few miles away from Oregon and Nevada state lines. "They built houses and took away the places to ride. That's why I had to come up here."

Norma still rides. Less than five feet tall and needing some WD-40 substitute for her aging body, she uses a mounting stool to climb aboard Sox. "Getting my leg up is more difficult than when I was 20," she confesses. "I don't gallop around like I used to. But at least I ride."

She wasn't selective as a young girl growing up on a ranch near Stockton, Calif. Her parents, Nell and William Thompson, had workhorses. "All the horses were black. I got tired of black."

Norma learned to ride western saddle, then decided to go bareback. She caught the eye of Jim McCleave, a Stockton horse trainer, and by age nine she was riding for him, competing as an amateur in jumping contests, including

Norma still rides. Less than five feet tall and needing some WD-40 substitute for her aging body, she uses a mounting stool to climb aboard Sox. "Getting my leg up is more difficult than when I was 20," she confesses. "I don't gallop around like I used to. But at least I ride."

CLOCKWISE FROM TOP: A teenage Norma flies over a hurdle. ➤ She was horseback as soon as she could walk. ➤ Norma's living room is decorated with old photos, trophies and awards from her riding. ➤ Norma looking spiffy with one of her horses. ➤ Sox, her current horse, begs for a treat.

© LEE JUILLERAT

PHOTOS COURTESY NORMA PETERS

the 1939 World's Fair in San Francisco. She later broke, trained and competed on her own horses. She rode English on Mister Mouse, aka Mickey, and western on Brandy. Both were quarter horses.

It took some work, but soon they were winning lots of medals. Norma kept only first- and second-place ribbons and awards and tossed out the others—enough to fill two large garbage bags.

Her riding days stalled after meeting Richard Peters, whom she later married. During World War II, when her husband was landing on Omaha Beach during the D-day invasion, most competitions were suspended. By then she was busy raising their son and daughter.

She paired up with Sox in 1998, which proved challenging because he had been abused by a previous owner. "I stuck with him and taught him to behave," she says, with a sparkle in her eyes. Over the years, the two have added to her trophy haul in trail competitions and parades.

After her husband died in 2002, Norma and Sox moved north from Stockton's encroaching suburbia and just a mile from her

daughter and son-in-law, Carole and Stuart Benner, and their family. Her son, Richard, is a retired doctor living in Michigan.

She's comfortable in Fort Bidwell and Modoc County's Surprise Valley because it's mostly populated with ranchers and horse people. When the weather's right, she and Sox head out the corral gate and travel around town, but not very fast.

"When I ride around the neighborhood," Norma says with a happy laugh, "the neighbors come out and chat."—*Lee Juillerat*

Me and My Agent

© Robin Payne

Marcus Uhart and Boo at the
Flying M Ranch in Yerington, Nevada.

Dee Kunzler, 92

A pretty hard looker.

"When I was born, Mom brought me on the train from Brigham City to Kelton and Dad met us with a car to take us on home. I'd got chicken pox when I was 10 days old. I was a pretty hard looker. A neighbor said, 'A kid that ugly can't live!' I fooled her!"

Born Dec. 25, 1924, to Harold and Vinnie Kunzler, Dee grew up on their ranch in Rosette in northwestern Utah with two younger sisters. When he was five, he rode the derrick horse for his grandpa. "I earned $15 that summer!"

Dee loves horses. His grandpa, Jacob Kunzler, gave him his first horse, a little buckskin filly. "She was a hard looker too, but raised some pretty good colts."

The first time at the wheel of a car, he drove it through the opened garage doors yelling "Whoa" and hit a trailer in the back. "I didn't like drivin' the old Model-A Ford."

When he was 10, Dee and his dad rode over the Raft River Mountains to find a mare. They stopped at Tracy's in Stanrod to share a meal, then went on to Yost where they found the mare at Teeter's. "Dad had a big old horse, Sandy. He choked down the mare and put a halter and saddle on, put me on her with a quirt, and sent me on my way."

They rode up Johnson Creek where the Indians camped and back over the mountain. "I was a pooped pup when we got home."

Dee can tell you stories about the good old days, like loading the last bunch of cows hauled on the Kelton train. He, Max Kunzler, and Al James drove them about 19 miles from Park Valley to Kelton. After corralling them, they were helping load cows for a neighbor on a Miller's Blue Ribbon Beef truck. The train pulled up and the engineer blew the whistle. "Them cattle just went swoosh and took the fence out and laid it over on Miller's truck. I remember Al James went up and give that ole engineer hell for blowin' the whistle."

Once Dee skipped school for fall roundup in Grouse Creek without telling his folks. He says: "Dad suspected that was where I went. Mom was madder than hell. That was quite an adventure for a kid."

After graduating from Box Elder High in 1944, he enlisted in the Army. His ship was headed to Japan when the bomb was dropped on

PHOTOS COURTESY DEE KUNZLER FAMILY

CLOCKWISE FROM ABOVE:
Dee (in center) with son and grandsons after a day of weaning calves in 2004. From left: Nick, Will, Dee, Del and Bradley. ➤ Dee watches Del cut meadow hay in Rosette in June 2016. ➤ Dee takes a break from shoeing to hold Del on "Bally" in Rosette, 1957. ➤ Dee's 90th birthday dinner at Morey's Steakhouse in Burley, Idaho, with daughters, Cheryl Leak and Marilyn Moncur, and son, Del Dee, in December 2014. ➤ Dee and his fiancé, Melva, in Brigham City, Utah, during furlough after basic training in 1944.

> **"Dad choked down the mare and put a halter and saddle on, put me on her with a quirt, and sent me on my way."**

Hiroshima, so they were rerouted to the Philippines.

He returned home and married his high school sweetheart, Melva Tracy, from Yost, Utah, on July 10, 1946. They called her the five-horse woman. Dee says, "I sold five of my best horses to buy her ring."

They lived in his grandparents' home and ranched with his parents, raising two daughters, Cheryl and Marilyn, and one son, Del Dee, along with Hereford cattle. Dee was awarded Rancher of the Year in 1979, the Rancher's Legacy Award from Utah Cattlemen's Association in 2008, and Box Elder County's Cowboy of the Day in 2009. He was an assistant brand inspector for more than 27 years. Today the ranch has mostly crossbred Angus cattle, and Dee helps Del's family, the fifth generation to work the ranch.

Dee rode in marathon horse races for several years. "We had a lot of fun. We set the fastest time the race was ever run from Burley, Idaho, to Oakley."

He also loves to hunt and has taken many neighborhood kids hunting. He tells the grandkids, "You haven't hunted till you shoot a box of shells, 'cuz once you shoot a deer, the fun's over!"

It's been a few years since Dee got on a horse, but he still supervises the ranch from the cab of his pickup. Melva passed away in 2008, but Dee is still going strong. "There's nothin' better than a good beefsteak," he chuckles. "Chicken will kill ya!"

—*Kellie Kunzler (daughter-in-law)*

Marie Ormachea Sherman, 83

`Cold mornings, dark nights.`

Winter mornings seem colder when she rolls out at four o'clock to drive the feed wagon, and summer nights seem darker when she's out irrigating. But beyond these two observations, Marie Ormachea Sherman finds ranch life as enjoyable as ever.

She's been at it all her life, with her earliest memories going back to age five. "I raised bummer lambs," Marie says. "My dad was a hard worker. He expected us to work hard and it didn't hurt us. We had to milk the cows and feed the lambs before we went to school."

Hard work was all her father, Tommy Ormachea, knew from the time he arrived from Alvia, Spain, in 1915 with 10 dollars in his pocket. He worked on ranches around Austin, Nev., scraping together enough money to buy a few lambs and start his own livestock business. He married a ranch girl from Fallon, Marguerite Kallenbach, and over the years they raised cattle and sheep on the Alpine Ranch about 50 miles east of Fallon and on the Jersey Valley, Cottonwood, and Home Station ranches in the Battle Mountain area. In the early '60s, the family took over the "home ranch" near Fallon from the Kallenbachs. Marie lives and works there yet, and in 2008 it celebrated a century of livestock production.

Her fondest memories are from the remote Alpine Ranch. "I just loved it out there," she says. "We were away from everything. We were kind of wild when we were little, kind of scared of people." Fear or not, among the people she met was Ray Sherman, a mine worker. They married and raised two daughters, Tomasa "Tammy" Ray, born in 1962, and Virginia Marie, born in 1965. A hired babysitter often watched the girls while

"I was camp tender for the sheepherders and cowboys. I would take them bread, meat, potatoes, string beans, corn, coffee, eggs, fruit, bacon, canned milk, and wine. I would put the eggs in the grain for the horses so they wouldn't break."

PHOTOS COURTESY MARIE SHERMAN

Clockwise from top: Marie Ormachea Sherman in 2013, still tending lambs. ➤Marie milking in 1949. ➤Clockwise from Marie are ranch helpers Nora Hunt-Lee, grandson Flint Lee, daughter Tammy Lee, son-in-law Kenny Lee, and grandson Tommy Lee. ➤Marie tends bummer lambs in 1943. ➤Horses were constant companions. Marie (on left) with friend among the remuda, around 1949.

Marie worked on the ranch.

Among her assignments was camp tending for the sheepherders and cowboys. They were so scattered and distant that it took an entire day to reach a camp and return. Each worker got a visit every five days. "I would take them

bread, meat, potatoes, string beans, corn, coffee, eggs, fruit, bacon, canned milk, and wine. I would put the eggs in the grain for the horses so they wouldn't break."

When not riding to the camps, Marie spent her days packing the supplies, putting up hay, irrigating, and feeding cows and sheep. And cooking. Marie baked tons of breadstuff over the years and is still known for her biscuits.

From their marriage until his death in 1987, Ray worked "out" most of the time as a logger and miner in addition to other jobs. "Ray was an amazing man with a mechanical mind," Marie says. "He could build anything. In the corral where I keep calves there are still feeders that Ray built."

Marie worked for her father until he deeded her the home ranch in 1973. She still works it today, with her daughter Tammy Lee and her family (husband, Kenny, and sons, Tommy and Flint) handling most of the heavy work. But she still likes being around the cows and sheep and keeps a handful of "nurse cows" nearby for raising leppy calves and bummer lambs. She also saves antique implements and everyday items and has amassed an impressive collection, the pride of which is a sheep-camp wagon.

Surviving on a ranch has always been a challenge. "Someone, whether it's the BLM or the water district," Marie says, "is always telling you what to do and how to do it."

The biggest change over the years, she notes, is the speed of life. "Back then, people worked harder and longer. But we weren't always in a hurry."

—Nora Hunt-Lee and Rod Miller

Careful Stewards

From riches to poverty and finally success, the Warnocks have ranched in Oregon since 1879. By Dan Warnock

The Warnock family had moved to an abandoned homestead in Dickison County, Kansas. Only one month later the famous Chapman Creek flood—the highest water ever seen in Cheever Township—would change the course of all their lives. During one night in late June 1869, the deluge came down so fast that Mother Nancy and her nine kids, the youngest under a year old, abandoned their "soddy" home with only a large quilt for protection. Her husband of 18 years, William Perry Warnock, had driven some cattle to market in Abilene and was on his way home. His horse made it across the flooded river that night, but he did not. Big brother, Levi, helped Nancy settle the estate, and the family moved to Abilene.

At that time, Abilene was the end of the railroad coming from the east. It was also the terminus of the Chisholm Trail. With all the railroad workers and trail drovers in town, Mother Nancy decided that opening her home as a boardinghouse would be a pretty good business venture. Her older boys night-herded cows for the drovers waiting to ship their cattle east. The rest of the kids took a team and wagon out on the prairie and gathered up buffalo bones, getting $3.50 for a wagonload.

Ten years went by, Mother Nancy remarried, and the boys wanted to go west. So in 1879 they all joined a wagon train and headed over the Oregon Trail to present-day Wallowa County in the far northeast corner of Oregon. My grandfather, Daniel "Big Dan" Welbaum Warnock, was the middle child of the nine and would have been 20 when they arrived in Oregon. That first year was spent near present-day Enterprise. The boys worked for

Big Dan was especially successful at raising horses. He loved thoroughbreds—the faster the better— and had a reputation for winning many races. Big Dan also raised draft horses and mules, selling lots of all three to the U.S. Cavalry Remount during and after World War I.

neighbors and split 10-foot fence rails that they traded for beef or sold for one-and-a-half cents each.

That first winter they lived on beef and rutabagas. Big Dan and his older brother, Uncle Dick, took up adjoining homesteads on Crow Creek, and most of the family moved there after brothers Tom and Alex bought a mixed herd of shorthorn and Hereford cows. (A two-pound roll of butter was bringing 35 cents in the larger towns, so why not milk that

bunch of cows?) Later, Tom and Alex headed down to the Snake River to work mining claims. All the provisions they needed for that first winter were loaded onto several pack-horses. Sadly, the animal carrying the whiskey didn't make it around a switchback. That caused a dry winter for Tom and Alex and the nearby creek forever since has been named Temperance Creek.

Mother Nancy died in 1883, the same year Big Dan married Mariah Mortenson, and the boys divided up the cows. Big Dan stayed on the Crow Creek ranch and Tom and Alex took their cattle to Temperance Creek. The brothers usually ran their cows together, using the Snake River with its many tributary creeks in Hells Canyon for winter feed.

Big Dan and Mariah raised a family of nine surviving offspring while building a cattle and horse operation. Dan expanded the Crow Creek ranch to 2,500 acres, rented pasture from neighbors and eventually wintered as many as 1,000 head in the canyons. Other homesteaders were moving in and in 1901 the Warnocks made their final drive out of Hells Canyon. My dad, Danny, was born in 1907. He was Big Dan and Mariah's last child, born 24 years after they married.

Big Dan was especially successful at raising horses. He loved thoroughbreds—the faster the better—and had a reputation for winning many races. Big Dan also raised draft horses and mules, selling lots of all three to the U.S. Cavalry Remount during and after World War I.

The riches were not to last, however. Big Dan bought into the First National Bank of Joseph, and in 1920 became its president. But by then war production of all foodstuffs was not needed. Mismanagement and a nearly 50 percent drop in cattle prices caused the bank to fail. Granddad and a few other stockholders tried to save it by liquidating personal assets. It didn't work, and in December 1923 Daniel W. Warnock declared bankruptcy to the tune of $108,191. Danny, the only sibling left at home, was still in high school.

Big Dan managed a hotel in Wallowa for a

After a 23-year struggle, Danny acquired this starter ranch in Sumpter Valley, Baker County, in 1946. Presently, our cattle are provided with year-round forage, from summer irrigated pasture and U.S. Forest Service permitted range to native winter range in Wasco County. With each new generation (the fourth is in charge now), the management of both grass and livestock has improved.

PHOTOS COURTESY WARNOCK FAMILY

while, but after his health failed, they decided to migrate west to the milder Willamette Valley. Danny agreed to go and help try to make a new start at ranching.

Danny's 20-year struggle through the Great Depression is a unique and sometimes amusing story. He milked cows, raised sheep, hogs and beef cattle, raised and butchered veal calves, worked at a creamery, and traded equities in real estate. Big Dan died in 1929. His final wish to go back to Wallowa country was honored and he is buried at Prairie Creek Cemetery east of Joseph, next to Mariah and near Mother Nancy. Danny married that same year, raised two boys, and was finally able to obtain a small ranch in eastern Oregon by 1946. He lived another 57 years.

Four generations, with a fifth coming on, have stayed and built on this beginning. Presently, Warnock Ranches Inc. is a family corporation operating in Baker and Wasco counties in Oregon, producing beef for a niche market. I like to believe that Mother Nancy and her son Big Dan would be proud. ∎

The third Dan Warnock was born in 1930 and lives with his wife, Jo, at the ranch headquarters in Sumpter Valley. His book, "You Can't Borrow Yourself Rich," is available for $20 plus $3.50 shipping from Betty's Books, 1813 Main St., Baker City, OR 97814, 888-202-6657 or bettysbk@bkrv.net.

CLOCKWISE FROM TOP LEFT: *William Perry Warnock holds my granddad, "Big Dan" (left), and his sister Mary in 1864.* ➤ *My great-grandmother, Nancy Warnock, holds Ruhama and Mary in 1864.* ➤ *Danny and his wife, Alice, on their 50th wedding anniversary on June 30, 1979.* ➤ *Big Dan in Joseph, Ore., ca. 1920.* ➤ *Danny (wearing tie) and all but one of Big Dan and Mariah's offspring in 1948.* ➤ *My dad, Danny, about age five.*

A Cowboy Romance

Bill and Nita Lowry, horseback early in the morning at the LU Ranch in Oregon.
By Carolyn Dufurrena

Bill Lowry grew up in the 1920s on a small high-desert ranch outside of Paulina in central Oregon. He had been "working out" for the neighbors since the tender age of "oh, about 11," he chuckles. By the time he was 17, he was buckarooing full time for the Stearns Cattle Company Triangle Outfit on the Crooked River. In the summer of 1941, on the cattle drive to summer country at LaPine, the buckaroos came in to eat at the main ranch cookhouse. A new girl, Nita Hein, was pouring coffee. She remembers that moment. "I was really shy. There was one buckaroo came in, and he kept staring at me. His green eyes were just shining," Nita says with a little twinkle.

Later, Bill would tell a story from that summer about the cooks doing "something with chickens" that involved a lot of squawking. He slipped around the corner of the bunkhouse in time to see that cute girl from town, axe in hand, horrified, with a headless chicken flopping between her brown-and-white saddle shoes, spraying blood all over. "I thought she was pretty cute before that," he says. "But after seeing how cold-blooded she was..."

He managed to recover from the shock and invited Nita along horseback one Sunday afternoon as he moved cattle. Nita says, "It was my first time pushing cows and I loved it." That December, the Japanese bombed Pearl Harbor. When Bill left for the war, he told her, "Now don't go getting married before I get back."

"I was only 15 then, and I thought, oh yeah," says Nita. But wait she did. The young couple stayed in touch for three wartime years. While Bill was on tiny Ascension Island in the Atlantic Ocean, Nita was finishing high school and cooking part time at the ranch for five dollars a day. Bill came back on a 30-day furlough in July 1945 to help with haying. He and Nita slipped away to Vancouver, Wash.,

on his third day back and got married. A week later, the United States bombed Hiroshima, then Nagasaki. The war was over. By Christmas, Bill and Nita would start their life together.

"When I got out of the Army, the ranch was paying a married man $150 a month. That was unbelievable," Bill says, shaking his head. "We lived on $50 a month and one tank of gas." Nita helped buckaroo and mowed lawns for the ranch. They ate at the cookhouse. They had all they needed and saved every penny because they had a dream: one day they would have their own place.

PHOTOS COURTESY LOWRY FAMILY

LEFT: Bill, just before his discharge in 1945. RIGHT: Newly married Nita in 1945. OPPOSITE: Bill and Nita share an intimate moment at the barn in 2010.

Harry Stearns needed help that summer. The wives went along on the drive to the summer country, sorting and scattering pairs along the Deschutes River on the way to summer pasture. Nita and Bill became true partners: he took care of the cattle; she'd ride when she could, cook and take care of camp. Over the next years, the family grew by two sons, Tim and Mike, who grew up horseback between the home place in winter and the Findley Ranch in summer.

"It was something you just couldn't wait to do," Tim remembers. "It was fun, it was tir-

ing, but you wanted to impress the old guys. When you started riding with the ranch crew, you started doing a whole day's work."

Bill became buckaroo boss, then eventually ran the whole outfit for Harry and his tribe, including the haying. The kids all worked together. They were family.

When the boys were old enough for school, Nita took Tim and Mike back to Prineville during the week, but the outfit still needed their help. One fall weekend the ranch was getting ready to ship steers from several places on the same day. Stearns asked Nita and the two young boys to gather one of the places along the river.

It was 60 miles from school to the Deschutes. They left school a little early that Friday afternoon, drove out to the ranch and caught their horses. Ninety-four head of wild steers hid in willow-choked sloughs and thickets along the river. Nita and the boys gathered quietly in the October evening, in and out of bog holes, wet meadow and thick jack-pine forest. They crossed the road in starlight, and Nita recalls that to keep the steers from spooking and running, no one said a word. They closed the gate on 93 head.

Bill and Nita finally saved enough to start looking around for their own ranch. By Christmas of 1965 they had their dream place, the LU Ranch east of Jordan Valley. Tim was in college by then, with Mike in high school. They built the place with the Hereford philosophy that they'd grown up with on the Triangle, kept saving according to Bill's lean-and-mean budget, and the ranch grew accordingly, to about 4,400 deeded acres, with accompanying BLM and state grazing permits. Tim married Rosemary Obieta and they came back to work the ranch and raise their young family. Mike went off buckarooing to Nevada, where he eventually settled down.

McQuaid, Bedford & Van Zandt in San Francisco and from Stewards of the Range, took up the banner, not only for the family, but for all ranchers in the West who depend on water rights being defined as private property. Stewards of the Range and countless individuals helped financially as the case dragged on. Tim read briefs and dug through boxes of reports and memoranda, while Bill and Nita kept after the ranch.

At the end of the day, at the end of 10 years, at the end of a million dollars in attorney fees, the case had gone all the way to the Idaho Supreme Court, which found that the U.S. government had no business taking the Lowrys' water rights. It went so far as to specify that "the argument of the United States reflects a misunderstanding of water law." It was a landmark decision, but the Idaho Supreme Court was unwilling to invoke the Equal Access to Justice Act, which awards court costs to the winner of a legal decision against the United States. It declined to award attorney fees to the Lowry family. So the Lowrys won, but at tremendous cost. The Idaho Supreme Court's decision conflicts with a ruling by the Nevada Supreme Court, which used the same precedents to award court costs to ranchers. Logically, the U.S. Supreme Court should resolve the issue. However, it refused to hear the case. Undeterred, however, the Lowrys persevere. Next step? The legislative route. Attorneys will petition Idaho legislators to propose that Congress amend the Equal Access to Justice Act.

The family, Tim acknowledges, couldn't have managed without the commitment—and patience—of their attorneys, as well as friends like Stewards of the Range and artist Jack Swanson. "People have stuck with us through the whole thing. We put something together, and we pledged to those people that we would carry this through. To the bitter end."

Bill and Nita have been through a lot in the last 15 years. But in spite of everything, each day brings this buckaroo couple joy. Nita loves the beauty of her place, "the independence of being our own boss," and the opportunity to work each day with her husband, "even if once in a while we do get into it," she smiles. Eighty-nine-year-old Bill still loves the days that he can buckaroo. "When you take off horseback early in the morning, the sun barely coming up, you're just at your ease," he says. ∎

Carolyn Dufurrena lives in Denio, Nevada.

At the end of the day, at the end of 10 years and a million dollars, the case had gone all the way to the Idaho Supreme Court, which ruled the U.S. government had no business taking the Lowrys' water rights...but declined to award attorney fees to the family.

Pressure started to build from environmental groups in the late 1980s to remove cattle from the open range. The LU Ranch summer country is in what is now the North Fork of the Owyhee Wilderness Study Area, and although language in the original wilderness bill specifically allows cattle to graze in wilderness, environmental groups wanted their cattle elsewhere. But people who don't need much, people who know what their freedom is worth, don't move easily. Bill and Nita stood their ground. With son Tim acting as point man, they persevered through year after year of mystifying grazing decisions by the BLM and under pressure from Jon Marvel and his Western Watersheds Project. Cuts to their permits resulted in an eventual 70 percent decrease in the value of the family's grazing permit.

Under pressure from the environmental lobby, agents for the BLM filed on the Lowrys' family water rights in the mid-'90s. These water rights, which are allocated by the state, had been attached to the ranch for decades. In addition, water rights are designated as being of "beneficial use," for which the federal government cannot qualify. Without these water rights, the ranch would not be viable. At first the Lowrys didn't realize they needed to lawyer up, but they found out soon enough. Tim once again became the point man and with the help of attorneys from

Norma Hapgood, 91

Avoiding the house.

Norma Hapgood has been good at many things, from driving farm equipment to growing a garden to moving cows on horseback. But some things have never come naturally.

"I am not a cook," she says. "'Bout the first breakfast I cooked was after Hillard passed away in 1995. No, no, being indoors, that's not me. I avoided the house. I always rode with my dad. I didn't like inside stuff."

For 91 years—"I don't know where that time went," she says—Surprise Valley in far northeastern California has been her home. Her parents, Roy and Vida Hanks, had a ranch outside Fort Bidwell, a small community near Lake City, where she's lived since 1949.

"We had a little starve-out place," she says of the family ranch, where they ran cattle and had milk cows. The fourth of seven children, Norma helped with all the chores. "I'll tell you what. I know how to milk a cow." She especially thrived horseback. "I'm sure I rode more than a million miles. I've loved every minute of it. I always had a saddle horse, but I didn't always have a saddle."

She and her sisters and brothers rode horses to school. "That was the fun thing in my life. My horses were tough, tough horses."

Her mother died when Norma was 12. "Life changed after that I'll tell you."

Her growing-up years included playing baseball as part of the Fort Bidwell town team, which included boys and girls. But more than baseball, Norma likes dancing. She found a

dance partner after graduating in 1942 from Surprise Valley High School in Cedarville and working at the grocery store. One of the store's customers was Hillard Hapgood, who visited from his family ranch in Calcutta, 40 miles away on the present-day Sheldon National Wildlife Refuge in Nevada. She doesn't remember the details, but is sure it was at dances where their romance blossomed.

"Hillard, he could dance," she says with a

PHOTOS COURTESY NORMA HAPGOOD

LEE JUILLERAT

"The snow was so deep it was at my stirrups. Talk about cold. We had icicles on our horses' noses."

"We had a little starve-out place. I always had a saddle horse, but I didn't always have a saddle."

CLOCKWISE FROM TOP: Part of Norma's daily routine includes feeding the chickens. ➤ *The Hapgoods gather firewood at their Calcutta, Nev., ranch. Their only source for cooking and heating was fires from juniper trees with logs cut to stove length.* ➤ *Norma horseback at the Lake City ranch in 1961.* ➤ *Norma as a child on her parents' ranch in Fort Bidwell.*

schoolgirl grin. They were married in 1944 and lived in Calcutta.

"We rode every place," she says, recalling cattle drives and the winter they gathered 40 head of pinto horses, drove them from the Sheldon to Cedarville, then over the Warner Mountains to Alturas. "The snow was so deep it was at my stirrups. Talk about cold. We had

icicles on our horses' noses."

Winters were spent at different places until 1949, when the Hapgoods bought the house that's still her home and added neighboring ranches. She estimates the family owned about 2,000 acres, ran 500 head of cattle, and raised hay on 500 acres. Her daughter and son-in-law, Bonnie and Joe Erquiaga, have taken over management of the ranch.

Her grandson, Jeff, a veterinarian in Colorado, provides the family with Red Angus bulls. "We're really proud of our cows," says Norma, who's also proud of her family. "Everybody knows how to work and everybody can do each other's jobs."

She's been hobbled by various surgeries. "I can't throw a rope," she moans. "But I get up at five o'clock and get to work." Asked if she rides, Norma answers, "Oh heavens, yes," although she needs help saddling a horse and steps on a bale of hay to mount. Even so, the perpetually smiling Norma has no complaints.

"Nobody had a better life than me. There's no part of ranching and my life I don't like. Made us a living. We worked hard. We were up early and worked late. Think of the fun times we had," Norma reminisces. "I've really liked every day of it. I'd like to go back and do it again."—*Lee Juillerat*

Coil "June" Redding, 91
A dauntless cowboy.

Coil Redding, known by all as June, knows a thing or two about cowboying, although his early life was much like those of many rural kids—doing what you had to do to make a living.

"I helped my granddad feed with a team," he says, "and my brothers, sisters and I all put up hay with teams. We fed some cows, and Grandma raised turkeys. We'd put those turkeys on a train. I remember getting our first tractor the summer of 1941. It was a John Deere H. It was small, but it pulled a combine. We combined grain for everyone around, earning two dollars per acre for cutting it. We also put up a lot of hay with that tractor."

The lure of cowboy life was strong. June worked for the Willcut family one fall, then for the Padlock Ranch in 1946, where he set up at the Conley Camp in the north end.

"At that time, the Padlock owned about 2,500 cattle, and we had to trail 300 head at a time from just inside the Montana line to the Conley Camp east of Crow Agency, 35 miles one way. I trailed cattle with some of those great cowboys like Barry Roberts, Little Barry Roberts, and Howard Drake."

During the roundup, the crew used a Dodge Power Wagon. "That's all we had. We'd pile everything into it. One time our cook was Bum Schubert; another time it was Billy Weeks. Mostly whoever was there did the cooking," says June. "When we branded, half the time we had a wood fire and half the time we had a propane tank. We'd brand about 2,000 head. At the Padlock we were horseback most of the time, gathering cattle and getting them worked. The ranch generally got us broke saddle horses. Sure, there might be a bronc or two, but we didn't pay any attention to them, we just rode them."

June married Jackie in 1951 and they lived at Wild Horse Camp on the Padlock. "We had two of our kids, Billie and Sam, and lived in a dugout. You had to go outside and go around

"At the Padlock we were horseback most of the time, gathering cattle and getting them worked. The ranch generally got us broke saddle horses. Sure, there might be a bronc or two, but we didn't pay any attention to them, we just rode them."

PHOTOS COURTESY REDDING FAMILY

to get to the second floor."

After leaving the Padlock, June did construction work to pay the bills. Then he went to work for J.B. Matson, who had come up from Texas to what was called the OW Ranch. In 1956, the Reddings purchased a ranch on Tullock Creek. June remembers them bringing three kids, a dog, and a beat-

up car. "I still own that ranch, 60 years later."

He had some heifer calves and got into the cow business. "We raised hay and alfalfa. We did everything we could to make a living so we could own this ranch," June admits. "I worked at the Holly Sugar Factory during beet harvest for 10 years. I bought a D-7 Cat and made money building reservoirs for local ranchers."

When he wasn't ranching, he was hitting the rodeo circuit, starting in the mid-1950s. "I rodeoed—bareback broncs, roping and bull-dogging—all over this part of the country and Wyoming." He did well, winning the all-around in

CLOCKWISE FROM TOP: June rides a saddle bronc at a rodeo in the early 1950s. ➤ June and Jackie at Montana's St. Xavier Rodeo in 1951. ➤ June ropes calves while branding at the Pitchfork Ranch in 2004. ➤ Proudly displaying his honor at the Montana Pro Rodeo Hall of Fame awards in January 2017.

Roundup, Mont., in 1953 and 1954. He rode a bronc in the 1977 Senior Rodeo in Roundup. In 2017, June and his family of three grown children and seven grandkids captured the 2017 Great American Rodeo Families Award.

The dauntless cowboy was still riding in the mid-2000s, roping and branding calves on his ranch and helping others as well. "I think ranching is as good a thing as anybody can do," June says. "You stay with it and take care of those cows and they will take care of you. Besides, we need people to keep raising cows. People gotta have meat!"—*Rebecca Colnar*

NEBRASKA · Spring 2020
Colleen Malmberg, 87
Middle sister on the prairie.

A gold and red sky flares in the east on the rolling prairie of eastern Nebraska. Bulging gray clouds to the west promise thunderstorms before long. The porch of Colleen Malmberg's trim log cabin faces that dawn, fringed by yellow maples. A little blue tractor is backed neatly into its spot at the edge of the driveway.

"Well, yes, I do drive it," she says. She's a little bit proud, at 87, of hooking onto a downed tree in the backyard and dragging it off. "But I don't use that tractor as much as I should."

A lofted central room is warmed by a crackling fire. The whole place smells of something sweet baking. Whip-thin, dark-eyed, practical to her marrow, Colleen is a farm girl. She drives a little Ford truck and is unafraid of the Nebraska winters.

Colleen is the third of five daughters descended from French Huguenots who eventually settled in the rich bottomlands of the Missouri River. The family ran to male offspring. Her great-grandfather, Nels DeLaschmutt, was a baseball player who raised nearly an entire team of boys. His son William was offered a contract with a visiting pro team one summer, but his dad, the coach, had hay to put up.

Colleen was born in the heart of the Depression. "We were poor, but we didn't know we were poor," she says. "Everybody was poor." The farm was 160 acres of rented ground four miles from the Missouri River, heated by a cob stove. "We pumped water. We cut ice from the river in the winter," she says. The sisters took baths in a galvanized tub.

"One-two-three-four-five, one after another, all in the same water," she remembers. They wore long brown stockings held up by garter belts and walked to the one-room school, where they comprised half the student population. "When our shoes got too short, Mom and Dad just cut the toes out, so we could wear them longer."

Like most everyone's daughters, they helped their mother in the garden and, like their sons, helped their father in the fields,

CLOCKWISE FROM TOP: *Truck full of babes in 1948. Colleen is hatless girl in truck on far left. The all-girl detasseling crew worked for Langren Seed Corn Company. She made $80, enough to buy school clothes.* ►*Colleen married Dale Malmberg in 1956 in Whiting.* ►*Colleen and her blue tractor. "We've had that tractor for 20 years," she says. "People keep trying to buy it but even if I get a condo in town, I think I'll just park it there."* ►*Missouri River flood, 1952. The farm was underwater for four months. Colleen's dad pilots a small boat headed for the house. She worked in Omaha with the Red Cross. "We took coffee out to the workers sandbagging the levees."* ►*The Reed sisters in 1940 with Boots. From left: Katherine, Colleen, Jeanne, Shirley and Donna.*

> **"I had a corncob on a wire and a coffee can with a little handle that was full of kerosene. I dipped the corncob in and lit the stalks on fire. When the corncob went dark, I just dipped it in the kerosene, and continued on down the row."**
> **Colleen was six years old.**

spreading out row by row to hoe the beans. Fall was the best time. Colleen's father would windrow the cornstalks after harvest, and the girls would burn the fields.

"I had a corncob on a wire," she says, "and a coffee can with a little handle." The coffee can was full of kerosene. "I dipped the corncob in and lit the stalks on fire. When the corncob went dark, I just dipped it in the kerosene, and continued on down the row." She was six years old.

By the time Colleen was 10, things started to get a little better. Their father was able to buy 100 acres of his own and keep the leased farm as well. The girls rode the bus to school in Whiting. "Katherine and I played basketball, and Jeanne was a cheerleader."

They helped their mother gather eggs, which they'd sell on Saturday afternoons in town, along with cream from the milk cows.

"We butchered hogs, but Dad would never butcher the cattle," she says. He wouldn't let the girls milk, either, "scared that we would dry the cows up. Well, our hands weren't that strong."

Later Colleen and Jeanne would take over some of the plowing "with a two-bottom plow" and an old Farmall M tractor. They detasseled corn in the summers on an all-girl crew when they got to high school. When the Missouri River flooded, which happened with some regularity, the girls moved in with relatives in Cedar Rapids. Her dad stayed to sandbag the house and keep an eye on the lake that was their farm until the water went down.

Colleen met Dale Malmberg, a farmer from Pender, Neb., at an old-time dance. Following her two elder sisters, who had also married farmers, they tied the knot in 1956 and moved back to his place. It wasn't all roses.

"I don't mind working at all," she says, a bit drily, but times for small family farms were never easy. The couple stuck with it for 30 years and four children, but after the crash in 1983, they sold out and moved to town.

These days Colleen, widowed but still intrepid, travels back and forth across the wide Missouri visiting her sisters. "We all live within 90 miles of each other."

There are still farms in the family, mostly run by grandchildren. For Colleen, the little blue tractor, the crackling fire in the fireplace, and dawn on the rolling Nebraska prairie will suffice.—*Carolyn Dufurrena*

Baby Cowboys at Work

© Robin Dell'Orto

I'll Be Right There!

Ketchum Wayne Dell'Orto, age two, at a Wooster branding in Copperopolis, California. He revels in everything cowboy and wants to be just like his daddy. He loves to throw a rope and ride Little Paint, a horse he shares with his tiny cousins.

Ready to Haze

Coney Elise Dell'Orto, 20 months, at the Dell'Orto ranch in Mokelumne Hill, California, assists Daddy haul cows. On this day they were hauling belled mother cows to Rail Road Flat before they were driven to the mountains for the summer. She is always ready to help move cows as long as she has the pink paddle.

The Grand Dame of Cowboy Poetry

Let the sound of laughter grace the day.
By Yvonne Hollenbeck

The year was 1929 when four-year-old Elizabeth Summers penned her first poem. The country was headed into the Great Depression and times were especially hard for farm families on the South Dakota prairie, but she constantly wrote verses noting the struggles as well as the good times experienced during her youth.

The whole family wrote poetry, leaving couplets around the house. Elizabeth's parents, John and Alma, and her three siblings lived a few miles from the North Dakota border, near the tiny farm community of Thunder Hawk, and their main form of entertainment each evening was listening to their mother or father read poetry or stories. At a time when there was no electricity, running water, or indoor plumbing, when life on the harsh prairies of western Dakota was extremely tough, little did we know that these humble beginnings would give rise to the nation's most beloved cowgirl poet.

Elizabeth was born Feb. 24, 1925, in the farm home of her parents, delivered by a doctor who arrived via horse and buggy. "I had a sister, four years older than I, who was very blond," said Elizabeth during a recent interview. "When I was born, I was brown as an Indian. My uncle told my parents, 'Well, you had a palomino; now you have a buckskin.' I was called Bucky for a lot of years."

In later years, she merely went by the nickname "Beth." She attended country school and graduated valedictorian at Thunder Hawk High School at 16. She attended the Black Hills Secretarial School in Rapid City, S.D., and spent a year at the University of Minnesota studying journalism and English, which was followed by employment as a bookkeeper for a bank in nearby Lemmon. She followed a friend to Washington, D.C., for one summer working as a waitress and was happy to return to her prairie homeland. She would often reflect on her childhood as a happy time, even though the family was never blessed with material wealth.

"Life was kinda fun though," she says. "We did more neighbor things. I rarely see my neighbors now." The family had an Overland car with a trailer behind when they left the small rented farm in search of jobs. Her father managed to find work as a mechanic at the Dorsey Garage in Philip, S.D., and her mother taught at Arrowhead School in Faith. They then worked for ranchers, eventually moving back to Thunder Hawk to another rented farm where they lived when Elizabeth attended high school. She reminisces on the difficulty her parents had trying to raise a crop during those Depression years when there was a shortage of moisture and a surplus of armyworms and grasshoppers. "My mother was a demon for schoolwork," Elizabeth recalls, "so I spent considerable time reading and studying. I intended to become a journalist, and then some idiot came home from the Army and I got married."

It was Nov. 8, 1945, when Elizabeth and a girlfriend attended a dance in Morristown, where she met S.J. Ebert, an area rancher's son who had returned home after service in World War II. They were married four months later and bought a half-section of land. They fixed up an old homestead shack where they spent their first 15 years of marriage and expanded their ranch and livestock herd. In 1962 they moved into a new farmhouse. The couple had three children: John, Jonni and Jayne.

Elizabeth kept pencil and paper available at all times and wrote poetry whenever an idea blossomed. Writing poetry was a way of venting joys and sorrows, marking a special event, describing the weather, or conjuring up a memory. She did not feel her poems were worth saving and many were tossed in the fire or trash can. It was not until 1989 that she

Elizabeth with her real cat and the toy cat her aunt made for her in 1931. LEFT: Elizabeth enjoys a spring day at her ranch home in 2017. OPPOSITE: Newlyweds Elizabeth and S.J. Ebert in 1946.

learned others enjoyed her talent.

Cowboy poetry, as we know it today, was in its infancy when a new cowboy poetry gathering was formed in Medora, N.D. Elizabeth was encouraged by SJ to attend. She was frightened to recite in public and it took much encouragement and arm twisting from her husband for her to relent. She was an immediate hit, not only with the attendees, but also with the other poets and performers, one of whom was the notable Baxter Black.

Shortly afterward Elizabeth was invited to perform at the National Cowboy Poetry Gathering in Elko, Nev. She credits her new friend and fan, Baxter Black, for the invitation. Elizabeth was extremely shy and nearly declined performing; if it weren't for the encouragement and support of Baxter, she would have never got on stage at this prestigious event. In fact, SJ had to drive around the block numerous times before Elizabeth agreed to make an appearance.

Although Elizabeth regularly wrote poetry, almost to her last days on earth, she discontinued public appearances in 2017. Health issues forced her to use supplemental oxygen, and she said, "I'm not going to perform as long as I have to drag that damned thing around." She found humor in nearly everything, including her own demise. Nearly a year ago, her family doctor advised her that she should get her things in order as she would probably have no more than two months to live. Eight months later, she scolded him for erring in his prediction.

Elizabeth suffered a broken hip and was transported to a hospital in Bismarck, N.D., for surgery. According to family members, she was alert and savvy, and when learning that the date was March 20, she quickly remarked, "This is our anniversary." Indeed it was and not long after that remark, with her family by her side, the Grand Dame of Cowboy Poetry took her final breath.

The poetry of Elizabeth Summers Ebert will undoubtedly be collected, recited, and cherished for a very long time. ■

Yvonne Hollenbeck, from Clearfield, South Dakota, is also a rancher and cowgirl poet of renown. She has performed all over the country for decades and often traveled with Elizabeth. They were absolute crowd pleasers. Their poems appear in the RANGE book, "Reflections of the West: Cowboy painters and poets."

NOTE: *Elizabeth Ebert dedicated her book, "Prairie Wife," to her husband on their 60th anniversary, March 20, 2006. A heading above this last poem in the book states, "Our youngest daughter has promised that when we die our ashes will be mixed together and scattered on this land that we love so well."*

When I Leave This Life
© Elizabeth Ebert

*When I leave this life as we all must do
And this prairie I've loved through the long,
     long years
There's a single boon that I ask of you,
Don't waste one precious day in tears.
Have a funeral if you feel you must
With the usual rituals for the dead,
A plain pine box, not satin-lined,
But with blanket, preferably in red.*

*No cloying masses of hothouse flowers,
Just a cluster of bright balloons, and then
No extolling of virtues I never had,
Just a simple prayer and a soft "Amen."
Let the memories be of the happy times,
Let the sound of laughter grace the day.
Find an old cowhand with an old guitar
To yodel me joyfully on my way.*

*And later, whenever the time seems right,
On a sunny day from a greening hill,
Scatter my ashes into the wind.
Then I shall be part of the prairie still.*

The Romance of the Range
© Elizabeth Ebert

*All the glamour of the cowboy,
And the romance of the range
That they show us in the movies,
I consider mighty strange.
Their nights are always balmy,
And their skies are full of stars,
And from somewhere in the background
Comes the strumming of guitars.
Her hair is long and curly,
And there's ruffles on her skirt,
And her boots are new and shiny
Without a speck of dirt.
His jaw is shaved. His clothes are clean
(I presume so is his mind)*

*And his horse is always trailing
Just about two steps behind.
He gently takes her trembling hands
In his own so big and strong,
And at the proper moment,
He bursts into a song.*

*Somehow while heading for our ranch,
We must have lost the trail,
For the glamour and the romance here,
By comparison, are pale.
We're out here in the cold and dark,
No moon or stars for light,
And we're bringing in a heifer
That is due to calve tonight.
My boots are pretty filthy,
And I just stepped in some more,
And my coveralls weren't purchased
At a Nieman-Marcus store.*

*And he's no movie cowboy.
He's looking plenty rough
With three days' growth of whiskers,
And a lip packed full of snuff.
We're tired and cross and dirty,
And we're chilled down to the bone.
There's no music in his cussing
…Just a steady monotone.
And 'though we're close together
And although our hands may touch
While we're sewing up that prolapse…
The magic won't be much!*

*So here's to cowboy glamour!
And here's to range romance!
We'd ride off in the sunset too
…But we never get the chance!*

Ode to Tofu
© Elizabeth Ebert

*The gentle cows upon our plains
That feed upon the grass,
And then, in turn, expel methane
In manner somewhat crass,
Are being blamed for making
Our atmosphere less dense.
They say someday we'll die because
Of bovine flatulence.*

*Does the answer lie in planting
Our rangelands all to soy?
If we abstain from eating beef
Will life be filled with joy?
Let's not accept this premise
'Til we check behind the scenes,
Just how much gas will people pass
When they're only eating beans?*

Basilio Susaeta, 92

War and peace.

"You boys would have been large men if you hadn't worked so hard so young," were the reflective words of Juan Susaeta after his two sons grew to manhood. However, Basilio held little regard for the injustice done to his build. "By the time I was 18," he says, "I was strong and confident enough to work any job!"

Born in Forua, Spain, on Nov. 26, 1926, Basilio entered a world recovering from World War I to Juan and Euleteria Susaeta. Second of seven children, his childhood wasn't yet marred by the effects of World War II. At seven years old, Basilio rose each morning to walk the mile and a half to Guernica to pick up bread, take breakfast to his working father, and go to school. Once school ended, he'd return home to help his mother cut grass for the cows, milk, and work in the family's lime mine. "Those hammerheads weighed 16 pounds!"

By the time he was 10, he could harness the oxen. Eventually Basilio and brother, Manuel, did the plowing. Basilio guided the plow and Manuel led the oxen. At the end of each furrow, it took both to trip the plow over to start the next row.

On market-day morning, April 26, 1937, when Basilio was 10, Juan watched unusual planes above Guernica. When a neighbor beckoned him to market, Juan declined, saying he didn't trust the planes. At two that afternoon, Germans bombed the crowded streets of Guernica. From their house on a hill, the Susaetas watched bombs explode

and fires rage. Immediately, the family escaped with their livestock.

"The road was full of people fleeing. Planes came in low and you could hear 'tat-tat-tat-tat' as they fired," Basilio recalls. People fell, their bodies abandoned as survivors fled.

> **Eventually, they returned to their farm, which had been ravaged by troops. A wheat crop was stolen from the field, the government to return for the straw. "There was nothing left. Not even the furniture." Persevering, the family rebuilt.**

PHOTOS COURTESY SUSAETA FAMILY

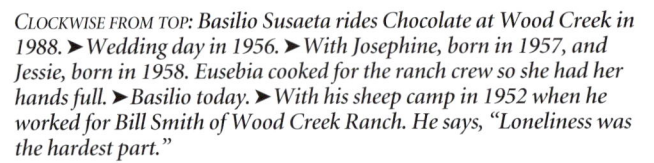

CLOCKWISE FROM TOP: Basilio Susaeta rides Chocolate at Wood Creek in 1988. ➤ Wedding day in 1956. ➤ With Josephine, born in 1957, and Jessie, born in 1958. Eusebia cooked for the ranch crew so she had her hands full. ➤ Basilio today. ➤ With his sheep camp in 1952 when he worked for Bill Smith of Wood Creek Ranch. He says, "Loneliness was the hardest part."

"My sisters rode in the wagon. They had to, or they would just cry at the bodies in the road."

As the battle raged on, now between Franco and Spanish government troops, they lived as refugees. Miraculously, none of the family was killed.

Eventually, they returned to their farm,

which had been ravaged by troops. A wheat crop was stolen from the field, the government to return for the straw. "There was nothing left. Not even the furniture." Persevering, the family rebuilt.

In 1952, after two years in the Spanish Cavalry and tired of working construction, 26-year-old Basilio came to America. "Loneliness was the hardest part," he says.

He worked in Idaho as a herdsman for Bill Smith of Wood Creek Ranch. A few years later, he met a Basque beauty working at The Basque Inn in Hailey. His herding dogs ran off, and while he never found them, he did find Eusebia. They married on Dec. 27, 1956, and remained at ranch headquarters in Grand View. With two girls—Josephine, born in 1957, and Jessie, born in 1958—and cooking for the ranch crew, Eusebia had her hands full. Basilio was ranch manager. In 1977, Bill sold the sheep and went full cow operation. For

Basilio, a knowledgeable cowman, it was a natural transition.

When Bill died in 1980, Basilio continued to manage. He retired in 1999, enjoyed his retirement party, and returned to the ranch the next day for work. Making daily trips to oversee things, he kept busy.

After 52 years, his beloved Eusebia passed away, and after 59 years, the ranch sold. Losing both left him lost for a time. Now he spends his days with his two grandsons, bouncing across empty desert in southern Idaho on cattle supplement trucks. If you ask him to describe his life, he thinks a moment, then says:

"It's been a long life. I've worked since I was a small boy. Ask anybody, war or work, I've seen it."—*Lyn Miller*

I Can Do It Alone!

© Skye Clark

Nevada Steadman, age five, climbs on Reata on a September day
to help her parents gather cattle near Georgetown, Idaho.

Merv McDonald, 94

Unbroken.

"I was born Sept. 28, 1923, on Fair Street in Petaluma. To a midwife." Merv McDonald is the second of four children born to Swiss-Irish farmers who had immigrated to California two generations before. He grew up on a farm in the foothills near Tomales.

Merv started ranching with 13 sheep "before I was 20 years old. I got married when I was 20." He and his wife, Dorothy Bettinelli, had $70, along with a flock of Suffolk ewes. To make ends meet he sheared sheep. "For eight years. I made 23 cents a head when I started, and when I finished I was making 47 cents."

In winter 1953, Merv did something he would rarely do again: he borrowed money. "I bought 800 ewes and moved up to Willits in Northern California. The snow come that winter. Runoff flooded the big lambing barn one night, and there were 300 dead lambs the next morning." They had drowned. "I sold $2,000 worth of lambs, $4,000 worth of wool, had $700 that year to raise my family."

Merv moved his family again, "trying to keep them sheep together and make it work."

Make it work they did until 1965, when he sold most of the flock and bought cows at Pierce Point from the McClures, old friends who had had the ranch since 1929. Pierce Point was one of the most beautiful saltwater ranches a person could imagine, property that had been bought by the National Park Service as part of Point Reyes National Seashore and leased back to the ranching families who had lived there since the 1860s. For 13 years, all went well.

"Everything was going fine, everybody was getting along, getting along with the park, and then boom! They say I got to move. They didn't want me there? Fine, help me get moved someplace. Don't just shove me out in the road. I wasn't fighting to try to stay there; I was fighting to have a place to land."

Merv traveled all over the West looking for a ranch that would "pencil out." There was a law that said they had to help him find a place to go. "We were in court, we appealed and we lost the appeal." Help was

In winter 1953, Merv did something he would rarely do again: he borrowed money. "I bought 800 ewes and moved up to Willits in Northern California. The snow come that winter. Runoff flooded the big lambing barn one night, and there were 300 dead lambs the next morning."

PHILIP FRADKIN PHOTO

PHOTOS COURTESY MCDONALD FAMILY

CLOCKWISE FROM TOP: Merv at Pierce Point Ranch just before eviction in 1978. ➤ Wedding photo, 1943. ➤ Merv and his daughter, Kathy McDonald Lucchesi, in the Petaluma ranch kitchen. ➤ Merv's driveway is lined with naked ladies which grow wild in the California foothills.

not forthcoming.

"The law said they cannot move you until the appeal was heard, which I understood." But a week before the court date, the park hired a trucker and a crew of cowboys to gather and ship the cows. Merv sold his mother cows, which were starting to calve, and moved again.

"They treated us just like the Bundys."

They bought 1,400 acres at Walker Creek near Petaluma in 1980. Over the years they leased more pasture from neighboring places; it adds up to almost 8,000 acres now.

Merv has been happy here. He lives in the old house with son Mike since Dorothy passed in 2001; daughter, Kathy, and son-in-law, Gino Lucchesi, live just up the hill. His son Bill makes saddles in Canby, Calif.

He still has a few Suffolk ewes. "My wife said, 'You better keep some of those sheep, that's what raised the kids,' and to this day I have 15 or 20 around." He takes care of a little bunch of heifers as well. "I went up that hill this morning on the four-wheeler, got off to open the gate and I could barely get back in the dang thing. I went all through those heifers, and I'd be from here to that cabinet [across the kitchen], and I couldn't tell if they were bagging up or not." But he still went.

At 94, Merv has never had a credit card. He brags about the "100 and some jars of fruit" in the basement, which he cans with Mike. He peels peaches, pears and apricots almost every day. "We put up fruit all summer," he says, "and we have fruit every morning all winter." He does physical therapy exercises since his stroke a couple of years ago, "every time I walk past this paper," prominently displayed on the kitchen counter. "I've seen a lot of change," he says, "but it's been a good life."

This is not a guy the world is ever going to conquer.—*Carolyn Dufurrena*

Celia Prentice Hall, 92

Things were always tight.

"The biggest blessing in my life has been my family and the lessons learned," says Celia Prentice Hall. At 92, she's spent 70 years on the Hall Ranch where she came as a young bride in 1949. "I'd gladly do it all over again if I could."

Celia Elaine was born on Dec. 15, 1927, in Yakima, Wash., to George and Murtle Prentice of Mountain Home, Idaho, while they were visiting her maternal grandparents.

"My father was a miner, but my grandparents came to Mountain Home to homestead. It was a difficult life." She attended school in Mountain Home. "Elementary school was wonderful. Full of music and art." She graduated in 1946.

For Celia, work started early. "When I was nine, I worked for a neighbor taking care of her little girl, the house, and the milk cow. I got a dollar a week, when she paid me." In high school, she found employment at the Ben Franklin store on Main Street. "The Haas family owned it. They were Jewish refugees who fled Nazi Germany." Along with a young girlfriend, she worked in the kitchen at the Mountain Home Airbase, called the Army Air Corps back then.

During World War II, Celia did her part in school. "When the war started, everyone rolled bandages for the Red Cross. We sold Victory Bonds, anything to help out." And while rationing affected everyone, for Celia's family it was nothing new. "Things were always tight for us."

Shortly after his return from service in WWII, Celia met Tom Hall through mutual friends. "Tom was home on June 13, 1947, and by June 21 we were going together." On Sept. 29, 1947, they married. Their first three years were spent on the Petan Ranch in northeastern Nevada. As an admitted town girl prior to ranch life, Celia had to adjust. "I learned to do everything from scratch. The hardest part was learning to ride a horse."

In 1949, returning to the Hall Ranch, Tom's home in Bruneau, Idaho, they welcomed their first child, Thomas Junior, in 1950, then their second, Charles, in 1951, into a home full of love but void of amenities. "I

hauled water and heated it on a woodstove. I washed laundry by hand and chopped wood to feed the stove."

It was work for the small family. "I traded two horses for a Jersey and a Guernsey we called Candy and Cake. We had those milk cows, pigs, and Tom's G.I. Bill. That's how we lived."

She learned the life, making cottage cheese, hard cheese, butter, the kids' clothes, planting a garden, and raising leppies. "Our old yard had no grass, so I went to the lower pasture and cut sod, then laid it out myself."

> **"When I was nine, I worked for a neighbor taking care of her little girl, the house, and the milk cow. I got a dollar a week, when she paid me."**

Celia ran a buckrake during haying, and in winter when Tom worked away from the ranch, she fed. "I learned to toughen up and I loved the work. Such a satisfaction comes from it."

In 1955, Tom and Celia bought the ranch from Tom's siblings. "It seemed every meal someone would stop by. I fed people for years."

In 1960, they welcomed a

PHOTOS COURTESY CELIA HALL

CLOCKWISE FROM TOP: Celia mows her hard-earned sod in 2018. ▶ The newlyweds at the Petan Ranch in 1947. ▶ At home on the Hall Ranch in Bruneau with Tom Junior (left) and Charles in 1955. ▶ Four-year-old Celia in 1932. ▶ Tom and Celia in 2010, close to a lifetime together.

baby girl, Mary. "Looking back, I should have gone out and played with the kids more."

After the kids grew up and Charles took over most of the ranch, Tom and Celia bought a little camper and toured ghost towns in Idaho and Nevada. Tom passed in 2011.

"I miss that man," she says. "The best part of being married was his sense of humor." Her advice is simple. "You've got to appreciate each other and appreciate every day." Her daily life has changed, but she still enjoys the simple things. "I love to read and I still get to live at home. My wonderful daughter takes good care of me."—*Lyn Miller*

100 Years

Pine Valley's Floyd Slagowski's productive century.
By Mary Branscomb

Tall, slender and craggy in his 100th year, Floyd Slagowski looks like what he is: a classic cowboy. Nowadays, though, he uses his gnarly fingers to braid bri-

He lives alone on Pony Creek in the house he and Charlene, his much-loved wife of 61 years, built of native rock about 40 years ago. It is a mile down the lane from

dle reins and bosals, not so much for riding. Recently he fell on his porch steps and broke a hip, so although he walks without a cane, he's decided not to saddle up any of several horses that graze just beyond the fence around his well-kept yard.

where son Dave lives in Nevada's beautiful Pine Valley, 50 miles southwest of Elko. Every morning Dave's wife, Billie Sue, comes up to Floyd's house to cook breakfast for him and Dave.

Charlene has been gone for five years.

Together they raised five children and a niece and nephew who all grew up listening to Charlene's music, eating her wonderful food, and appreciating her drawings. The two youngest sons continue to operate the Slagowski Ranch today. Son Carl and his wife, Sharon, live 10 miles down the road.

Always a meticulous man whose memory is detailed and exact, Floyd remembers visiting with Harry Webb, who rode broncs for Buffalo Bill and lived and ranched for a while from a cabin at the south end of Pine Valley.

"I stopped sometimes to visit Harry when I was gathering cattle for Ross Plummer," Floyd recalls. "The first time, I enjoyed a cool beer he kept in a nearby pond. Then I spent the night. I did not gather any more cattle that day because Harry was a good storyteller!" *[Harry's raw and real stories are shared in "Call of the Cow Country" and "The M Bar," found at rangemagazine.com.]*

With Charlene's encouragement, Floyd spent years researching and writing Pine Valley's only history. He has ridden every inch of the valley and its mountains. Sometimes he took grandsons along on camping trips to scatter salt or gather cattle.

Finally, in 2010 he self-published an accurate and illustrated history, a two-pound book titled "The Pine Valley Puzzle." It took him decades writing with a pencil on yellow legal pads to be typed later by his neighboring schoolteacher/rancher friend, Rita Stitzel. "I couldn't change a word," she says. "Floyd would notice everything and make me do it *his* way."

He served as a brand inspector for 17 years, was on the Mineral Hill and Eureka school boards, and was an officer for the Eureka Farm Bureau. He also received the Nevada Cattlemen's Association award for riding horseback for 100,000 miles.

Floyd was born in Wyoming near Henry's Fork, 50 miles south of Green River near Burnt Fork. "My father, Eli, had taken up a homestead when he was only 19 and married Susan Merchant the next year. They produced six boys and three girls while they 'proved up' the property."

Floyd was two in 1918 when his mother got sick and World War I ended. That year a devastating flood washed away Eli's house and outbuildings. Although he and his older sons rebuilt, Susan died in 1929, the stock market "broke," and the Great Depression hit.

Eli made sure his younger children got as much education as was available, so for two bitterly cold winters, Floyd hitched up and drove the school wagon through deep snow pulling his siblings and others to school. "When I turned 17, I and an older brother headed out looking for work."

They found several jobs in Wyoming making hay. That's when Floyd learned to enjoy handling draft horses. "We earned a dollar a day and board and another 25 cents if we wrangled horses. Another job entailed gathering, breaking and riding wild colts for a rancher who sold horses to the cavalry."

In 1936, Floyd spent a profitable year on a cattle ranch learning the cattle business. When his employer lost the ranch in the Depression, Floyd headed to Nevada in a seven-year-old Model A that used not much gas at 17 cents a gallon.

One of Floyd's experiences happened when he was down to four dollars in his

FROM TOP: Floyd rides Stormy at a branding on the JD Ranch in 1942. ➤ *Horses have been a major part of Floyd's life since he was a 13-year-old boy in Wyoming.* ➤ *Charlene and Floyd were happily married for 64 years.* ➤ *The "school bus" that Floyd drove in the winters of 1931-32 and 1932-33 in Wyoming. OPPOSITE: Floyd is happy, surrounded by treasures.*

pocket and needed a job—any job. Pete Laborda in Eureka told him about a rancher in Diamond Valley whose hired man had just quit. "Pete said a certain horse kept unloading him but when I got to the Saddler Ranch, I was happy to get a bed in the bunkhouse, a good meal, and the promise of $45 a month. My pay was $30 in Wyoming."

The next day Riney Saddler had put the "certain" big black horse in his string, the first to be ridden. Floyd led him to his car to saddle up. "I had been around enough broncs by then that I could tell he would sure enough buck, so I figured we'd just get it over with," he says. "Didn't even untrack him or pull up his head. Just drove in both spurs. He bucked right through the boys in the corral. When he was done, we started the day's work. I'd ridden horses that bucked harder. I kept that horse in my string and enjoyed him for the next few years I worked for Riney."

In 1942, Floyd was drafted. He spent most of his Army career in the 41st Quartermaster Pack Mule Troop based in Hawaii where he learned to shoe the Army way, taught other soldiers how to shoe horses, and packed mules up mountains with artillery and supplies for recruits preparing for the expected jungle warfare. Then he was sent to Fort Collins, Colo., and was in Denver when the United States dropped the first atom bomb. "I never saw such a celebration," he says. "All those soldiers knew that now we would not have to invade Japan."

Floyd was discharged in 1945 and returned to Pine Valley. He and Charlene were married in 1946. They built a ranch that spreads 35 miles from one end to the other and is 16 miles across. They run 1,000 head of cows on public and private land. Many of their grandchildren are in the livestock business one way or another and some grandsons are champion saddle-bronc riders. There are 13 Slagowski grandchildren, three great-grandchildren, and more on the way.

The ranches in Pine Valley have changed hands several times in his lifetime, but Floyd says they have become better managed. "They haven't been cut up, so there are not many more people. But now there is telephone service!" ■

Mary Branscomb lives in northeastern Nevada, 100 miles northeast of the Slagowski Ranch.

Richard Snider, 92
Sheep rancher and soul doctor.

Richard was born July 28, 1927, in Sheridan, Wyo., but spent his early years on a sugar beet farm in Leiter.

"For the first three years I attended grade school, I rode my Shetland pony, Queeny, there. I was a horse-crazy kid. My dad raised sugar beets using a team of mules to do his farming. Growing up I learned how to cuss and how to open gates."

When he was about eight years old his dad had a pickup truck with wooden spoke wheels and lights that didn't work. "He'd ask me to come along when he went to town so I could open the gates and drive him home. I'd sleep in the truck until he was done in town, then drive him home and get those six gates."

Richard's father had a few sheep and cows but times were tough. His grandparents had homesteaded in Clearmont, Wyo., "where nothing would grow but a little grass and rattlesnakes. There weren't many roads back then. There also wasn't water, lights, money or many cars, but there were lots of horses and, in the 1920s, a lot of moonshine in those hills!"

In 1939, the family moved to Sundance. "My sister and I rode horses five miles east to Bear Lodge School for seventh and eighth grade and boarded out in high school for the four years during the war. Everything was rationed."

Richard remembers the days of grasshoppers and Mormon crickets in 1936. "The sun was shining, the wheat looked great; then the sun disappeared and there were millions of grasshoppers. The grass had been three feet high and they ate all that grass in about a week. You couldn't find grass even a half-inch long."

Wyoming ranchers always talk about the winter of 1948-1949. Richard says they put all of their livestock into the outbuildings. "After the big storm we made trails so the livestock could get out. It took a week to get to the road with three horse-drawn teams taking turns breaking the trail. The drifts were 40 feet deep in some places. I remember we were trying to get hay on a tow rope out to 200 ewes. My horse tipped off into 40 feet of snow and we went out of sight. It's a wonder I didn't die."

Richard married Irene Roark on Sept. 15, 1946. "I remember I was standing in the high school auditorium. I didn't have a girlfriend. There was a girl standing there named Irene and she threw her arms around me and gave me a kiss. When I got out of school, we got married."

oped the place into a working ranch with 900 ewes, 200 yearlings and 200 yearling bucks along with 100 black cows and 95 calves. They ran on 5,000 acres between their land and a lease from a neighbor. During the height of the sheep business, they would have 15 sheepshearers in the shed. "The best shearers came from Australia. I got about 48 bags of wool, 200 pounds a bag. I made a lot of money on wool."

In addition, the Sniders farmed 200 acres of wheat and 100 acres of oats. "I've been

> "My grandparents had homesteaded in Clearmont, Wyoming, where nothing would grow but a little grass and rattlesnakes. There weren't many roads back then. There also wasn't water, lights, money or many cars, but there were lots of horses and, in the 1920s, a lot of moonshine in those hills!"

CLOCKWISE FROM TOP: A successful shearing with 48 bales of wool in 1964. ➤ Richard, Irene and baby Gary in 1948. Richard met her in the high school auditorium and he didn't have a girlfriend. "Irene threw her arms around me and gave me a kiss. When I got out of school, we got married." ➤ Preacher Richard saving souls at home. ➤ Irene and Richard on their 50th anniversary in 1996.

Irene and Richard bought the Snider home place in 1953 in Sundance with a loan from a neighbor, which they paid back. Everything was used or second-hand. They devel-

hailed out, droughted out, priced out. There was one year we sold wheat to India and got $7.50 a bushel—that was my one good year farming." They also raised two sons and a daughter. Irene passed away in summer 2019.

"I became a New Testament Christian when I was 32. If I went in to pay for gasoline, I kept preaching," Richard says. "I had a sign on my gate that read, 'This is where the Church of Christ is.' It's tough to get people to be faithful. Some of the people I talk to will tell me, 'If I see you again, it will be too soon.'"

—Rebecca Colnar

Al Pinetree, 80

Hammer and tongs in the High Sierra.

Four thousand feet above the Pacific Ocean the rough granite shoulders of the Sierra Nevada are shrouded in smoke. In one of those high valleys is the hamlet of Badger, Calif. Over one of those smoke-shrouded ridges lives Al Pinetree, a blacksmith and small-time rancher, and his wife, Sherrie.

Al was born on a dairy north of Trout Run, Pa., where he lived for a couple of years with his single mom. "I never knew my father," he says. Raised by his maternal grandparents, he started day working at 15, and by 17 had left home to join the Navy.

"It was 1962 or thereabouts," Al says. "It was before we got involved in Vietnam." He learned to be a mechanic in the service. "I knew if I could fix things, I'd never have trouble finding a job." He became a specialist in fixing things like paving machines and Caterpillar tractors—he fixed 20 for the Army—moving all around the country. Eventually he landed in Los Angeles working for a Cat® dealer with two Dutchmen, aircraft mechanics who were veterans of World War II. They called Al "Pinetree," a German translation of his adoptive surname. Later, he legally changed his name in their honor.

"All the jobs were in L.A.," Al says, "but one job was in the high country. I came to Badger to fix a tractor and I never really left."

There was good natural feed and plenty of pasture. "I knew about horses, had grown up around horses. A local horseshoer showed me a thing or two." Soon Al was making horseshoes. Not long after that, he was also nailing them on. "I could shoe horses other people couldn't. I'm only five-eight," he says, "so I could get under those big geldings."

He learned about shoeing the mules used in pack strings and logging in the mountains. "This is rocky country," he says, "and my shoes stayed on those mules."

Along the way, Al bought a small acreage and started a little bunch of cows with a Santa Gertrudis bull and a red heifer. "I wanted cattle with a little ear, bred for hot weather and bugs." He day worked and helped the neighbors. One of those neighbors hosted a series of brandings that lasted 10 or 12 days. Al says, "It was the biggest branding of the year." There was a woman there helping, "a nice

CLOCKWISE FROM LEFT: Al and Sherrie married 20 years ago on the hill above their house, accompanied by dogs and a few friends. He jokingly made a nose ring for himself of sweet iron. "It made her laugh." ➤ Al shapes a horseshoe. "I mostly make hearts out of horseshoes these days," he says, "and send them as a gift from my heart to people who make a difference in this world." ➤ The newlyweds wander home. "I got this old house, leaks a little bit. Been here 45 years." ➤ Neighborhood branding in the 1990s where Al met his bride. Here he stamps one on as a young friend from Georgia holds the heels.

PHOTOS COURTESY PINETREE FAMILY

"All the jobs were in L.A.," Al says, "but one job was in the high country. I came to Badger to fix a tractor and I never really left." He could have gone on disability 15 years ago, "but hell, I can still work."

kinda gal, but she always wore sunglasses. I never saw her without her sunglasses." Then some mutual friends set them up.

"I went inside and she was sitting there. I saw her eyes, and it was like an electric shock. 'That's the gal for me,' I said." At 60, Al had never been married. Sherrie was 39. She had two teenage sons; he had two grown daughters. After the boys grew up, he asked her, "Think we ought to get married?"

"I'd like that," she said. "We'll just go up there on the hill." That was 20 years ago.

Today Al's eldest daughter, Maria, is married to a logger and lives across the road. Daughter Sophia works in the Bay Area. Maria's three sons and Sophia's daughter are out on their own, and a great-granddaughter has graced the scene.

Al, at age 80, is mostly retired from "nailing 'em on," but he still tinkers around in the blacksmith shop. "I make a lot of stuff from old horseshoes, from rebar. I'm working on a mule-shoe hoof pick, with a bottle opener on the other end," he adds proudly. When the weather gets right, he says he might go around these fences too.

"I could have gone on disability 15 years ago, but hell, I can still work," he says. "We're not poor. I'd rather tell stories and make horseshoe hearts."—*Carolyn Dufurrena*

Ranching in Heaven

© Amber Oberly

CLOCKWISE FROM LEFT: *Anson, starting as a todddler, feeds his family's flock of chickens.* ➤ *Emmy (9) loves to help out. Here she feeds a bum calf and gives vaccinations to the dog named Pup.*

Whether it's feeding chickens, bottle-feeding an orphan calf, or making sure the dog has shots, Charles and Amber Oberly's ranch kids learn to help out on the place from a very young age.

In these photos, Anson and his sister, Emmy, take care of their chores a mere 25 dirt-road miles from Big Timber, Montana, playing at work and making sure that no chickens escape the yard.

Phyllis Phalen Griffin, 84

Always smiling.

Phyllis Griffin remembers the Great Depression well. "We ate rabbit, we had a pig so we had pork, and there were plenty of sage grouse and sage hens. Dad had a few cows. Unfortunately, we droughted out in 1937. Between the drought and grasshoppers there was no grass."

Instead of complaining about the scarcity of her early years, Phyllis smiles when she remembers the good times. "There are lots of good memories. My mom could make a meal out of anything, and I had so much fun with my siblings. It was a great childhood. We didn't have anything, but we didn't miss it. We were together, we had good parents, and we were never hungry."

She was born on Box Elder Creek in Ekalaka, Mont., on Sept. 21, 1929, just before the big market crash. "I was born in a homestead log house. My grandma and grandpa had lived there first and then, in 1921, my folks moved in. Four of us were born there with my grandma serving as midwife."

Phyllis and her brother gathered beer bottles and pop bottles to earn enough to go to a show. "If we got a case of beer bottles, we might be able to go to the show and have pennies for penny candy. We got a penny for beer bottles and two pennies for pop bottles. We had a little wagon to collect them. We would go to the back door of Wylie's bar and he always paid us well," she chuckles. The one thing she promised herself was at some point in time she would own 100 cows and have land to run them.

Phyllis met Buford Griffin in 1944. "I went to work for a ranch that summer earning 50 cents a day. The second year I worked at that ranch, I got a raise to a dollar a day," she says. "Buford was working for them then. I did everything. I'd wash clothes on the scrub board, I did some cooking, and I learned to ride and do all types of ranch work."

Phyllis and Buford were married in 1951 in Miles City. The new bride loaded her possessions into a couple of boxes. "We didn't have much. Buford went to work for Bud Brown near Ismay in 1950. We worked there earning $150 a month and Bud said we could run 20 cows."

She had an old wood cookstove, a cast-iron

PHOTOS COURTESY PHYLLIS GRIFFIN

CLOCKWISE FROM TOP: *Phyllis, fourth from the right, at branding camp in 1948. Buford is second from the right.* ➤ *Phyllis gets ready to can beets with her granddaughter, Nicole Rolf.* ➤ *With brother, Bill, in the early 1940s.* ➤ *The Phalen siblings in the early 1930s; Phyllis is on the horse behind her brother.* ➤ *Buford and Phyllis.*

skillet and a Betty Crocker cookbook. "That was about it. We didn't have running water," she remembers with a smile. "That's where I really learned to cook and other kitchen skills like canning beef. Beef was our prime staple."

The two worked side by side to fulfill their dream, with help from Brown. In 1955, they had the opportunity to buy half of his ranch, and once the first half was paid, they bought the rest.

As many people did in ranch country in the 1950s, the Griffins started with Herefords, then crossbred them with Angus. Today the Griffin Ranch breeds an Angus-Simmental cross. The couple has more than fulfilled Phyllis' dream of 100 cows and land to run them on.

> "My mom could make a meal out of anything, and I had so much fun with my siblings. It was a great childhood. We didn't have anything, but we didn't miss it."

"I wouldn't trade ranch life for anything," she says. "It provides a wonderful opportunity to keep your family together. We have meals together and talk about the day's work. The work ethics my three sons learned growing up on our ranch is worth its weight in gold. Two of our sons still work here and four out of our six grandchildren were raised here."

Phyllis still gardens. "I got my squash and potatoes planted already, and I'll plant tomatoes, cucumbers, carrots, cantaloupe and watermelon. I still help with cow work, too. We work together as a family. Having family is the best asset you can ever have."

—Rebecca Colnar

Tough Times at Wolf Creek

Surviving the Depression in Montana.
By Marie Poloson as told to her daughter, Grace Larson.

My brother Fred and I have crooked toes. Fred said his came from wearing a poor pair of shoes bought at the mercantile. During the Depression years at Wolf Creek, we wore those shoes a long time.

Mom made most of our clothes. Dad was making money at the Helena Valley Sheep Ranch. We had nice bedding and about everything we needed, but as we grew older we outgrew our clothes. Mom made jeans for Fred and Bert from the khaki riding outfits out of her trunk. She sewed everything by hand and taught me how to sew and embroider. When I was older I embroidered pillow cases and sold them at the Lonepine store.

"In those days there weren't any jobs so getting through the winter was difficult. People stole wood. They stole everything. Miss Mosier, who cooked for us, had so much wood stolen from her little house in Camas that she moved it into the house, covered it with her mattress, and used that for her bed."

We got our shoes from catalogs, the mercantile in Hot Springs, or the mercantile in Plains. The one in Hot Springs covered an entire block. It was destroyed by fire in 1931.

Men would come and work all summer if Dad would let them stay through the winter. So many of the men were alcoholics; they would stay sober for months, then go on a big drunk. Phillip Straus was one of the men who stayed at the ranch. Phillip drove us to school sometimes. Bill Murray took one vacation a year and spent it at the bar. He was a "reformed alcoholic." Dad talked about the time Bill hollered "whoopee," threw his hands up, and somersaulted off the barstool.

In those days there weren't any jobs, so getting through the winter was difficult. People stole everything. Miss Mosier, who cooked for us, had so much wood stolen from her lit-

ABOVE: Marie Poloson on Star, ca. 1938.
LEFT: Daughter Grace, age three, rides an old ewe in 1943.

tle house in Camas that she moved it into the house, covered it with her mattress, and used that for her bed.

During those poor Depression times people would turn their horses loose, and some even turned their cattle out. They ran wild in the hills. The horses usually survived the winter, but sometimes the snowdrifts were so deep the cattle couldn't get to the grass, and they died in the gullies.

During those poor Depression times people would turn their horses loose, and some even turned their cattle out. They ran wild in the hills. The horses usually survived the winter but some cattle died in the gullies. The snowdrifts were so deep the cattle couldn't get to the grass.

Dad always had enough hay for the livestock. Wool actually brought a fair price during the '30s and '40s, but sheep prices were awfully low, around a dollar a head. The weather was cold. Forty below was nothing and the snow was so deep it covered all the fences.

There wasn't any money, but we were really lucky because we had our milk cows, chickens, and hogs. Mom kept canned milk and

could butcher it. Dad would lower the hog into boiling water so us kids could scrape the hair off the hide. This made nice side pork with that wonderful rind to chew on.

Mom always had about 40 chickens, and she had ducks. Someone had given her a bunch of mallards which are generally wild. Mom thought it funny that these ducks

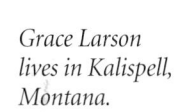

was pure. People used cheesecloth for netting over baby buggies so the baby would be safe from mosquitoes. Cheesecloth was even used for bandages because it washed so clean and dried fast.

During that time I earned some money by herding Indian Department cattle. I used that and what I made from bum lambs for school clothes. Mom grew up during hard times so she knew how to make do with what she had. That's what helped Dad get through the Depression. That and the fact the bank wasn't about to foreclose when sheep were a dollar a head. ■

FROM TOP: Jess Miles, horse trainer, on Night in 1948. ➤ The Poloson pack string heads for sheep camp in 1946. ➤ Dan dips sheep in 1936.

Grace Larson lives in Kalispell, Montana.

celery cool by setting it in cold water. We butchered and cut up our own meat. Mom stored vegetables, apples, and pears in the cellar. Mom canned a lot of our meat before we had the deep freeze in the cookhouse. She had a "cooler box" by the side of the house that kept meat frozen during the winter.

Dad would butcher a fat lamb or a wether, a castrated sheep a year or two old. Home-raised mutton was good, but Dad always said, "Don't let the wool touch the meat." When Dad skinned a sheep he would roll the hide back so it wouldn't touch the meat.

When we butchered our pigs we used a block and tackle tied to a horse. By leading the horse forward this raised the pig up so Dad

never laid eggs. The white ducks were laying all the time. About three weeks after Mom's comment the mallard ducks came down the hill from the buck pasture followed by lots of ducklings. Mom had close to 100 baby ducks, so we ate a lot of duck that year.

When we would buy cheese at the butcher shop it would be wrapped in cheesecloth. That is how cheesecloth got its name. We used a lot of it. Mom would cook pork fat and rinds until the fat melted; then she'd strain it through cheesecloth so the lard

Russell Wyatt, 94

Pursuing dreams.

Born, raised and continuing to live in South Dakota's Fall River County, Russell Wyatt has had many careers in his lifetime and, at the age of 94, still works. "I grew up on an irrigated farm that was homesteaded by my grandfather in 1881," says Russell. He and his two brothers worked alongside their father to raise produce for marketing. "I mowed and raked many acres of alfalfa with a five-foot mower and 10-foot dump rake pulled by a two-horse team," he says. "I was 10 when Dad purchased our first tractor in 1936. I was driving it by the time I was 11."

Russell rode a horse to a one-room school for seven years and graduated from high school in Colorado.

During the building of Angostura Dam near Hot Springs, S.D., Russell worked as a carpenter and truck driver. After the dam was completed, he worked for Peter Kiewit Construction to build the water delivery system for the Oral, S.D., irrigation project.

Russell married Oleta Harley in 1947. They had two children, Jerry and Peggy. They bought an irrigated farm on the Angostura Project in 1958 and built a new house. This was the first home he had with indoor plumbing. Russell ran 150 cows and 200 sheep on the farm all year and wintered 400 calves. "I rode a lot of horses but never in a rodeo," he quips. He raised sugar beets, dry beans, corn and alfalfa. Sadly, Oleta passed away in 1962.

On Easter Sunday in 1963, Russell married the local schoolteacher, Betty Thackerson. Russell and Betty had two more children,

Audrey and Patrick. Russell says, "I am still amazed at how well a city girl from Alabama adapted to be a rancher's wife in western South Dakota. She became a wife, a mother of two teenagers, and the cook for the haying crew the day she became my bride."

Looking to expand his income, Russell became a real estate broker in 1968. In 1972,

"I was 10 when Dad purchased our first tractor in 1936. I was driving it by the time I was 11."

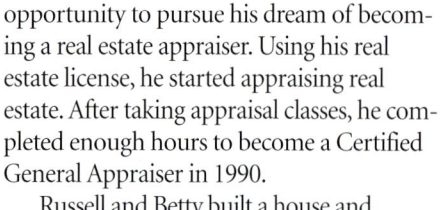

CLOCKWISE FROM ABOVE: Russell and Mom move into their homestead in 1928. ➤ Leaving home at two years old with bag packed. ➤ The happy couple at their present home in Hot Springs, S.D. ➤ Russell and little brother, Harold, with Oscar the turkey. ➤ At the homestead at age nine with dogs. ➤ Russell drives his favorite tractor. ➤ Russell, Betty and great-grandkids, Janna and Wyatt Curry.

PHOTOS COURTESY WYATT FAMILY

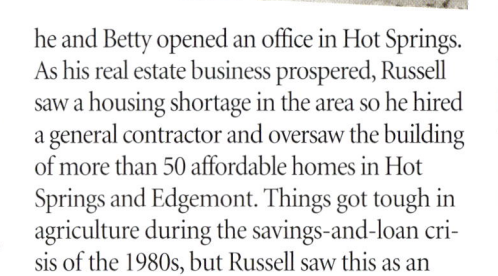

he and Betty opened an office in Hot Springs. As his real estate business prospered, Russell saw a housing shortage in the area so he hired a general contractor and oversaw the building of more than 50 affordable homes in Hot Springs and Edgemont. Things got tough in agriculture during the savings-and-loan crisis of the 1980s, but Russell saw this as an opportunity to pursue his dream of becoming a real estate appraiser. Using his real estate license, he started appraising real estate. After taking appraisal classes, he completed enough hours to become a Certified General Appraiser in 1990.

Russell and Betty built a house and moved to Hot Springs in 2005. He continues to work with his youngest son, Patrick, who took over the appraisal business in 2019. "I heard during the financial crisis that the two worst businesses one could have was ranching and real estate," Russell says. "I had both! We sure had to tighten our belts, but we made it."

Legacy is important to Russell and he's proud of the fact that he sold the farm to Peggy and her husband, Russ Sanders. They now operate the farm with Russell's grandson, Carl, and his wife, Kari, and their three kids, Megan, Kyle and Ray.

Russell says: "I was 65 years old when I got my appraiser's license. I couldn't type, had never been to college, and didn't know squat about computers! How did I ever make it to be an appraiser for almost 30 years? Today, I have a nice home on a small acreage, a loving wife, children, grandchildren, great-grands, and Betty and I are debt free. How blessed can I be? God is good!"

—*Audrey Shrive (daughter)*

Riding, Roping and Learning

© Kathy McCraine

Kreed Kasun, five, has been riding by himself since he was two, when his folks gave him an ornery paint pony. After the pony knocked him off on a gate post one too many times, he got up, dusted himself off, and announced: "That's it! No more little horses."

Kreed's dad, cowboss KJ Kasun, runs a tight ship on 134,000 acres of rangeland with 1,200 head of mother cows and stockers near Williams, Arizona. Kreed is the youngest of his junior cowhands, along with brothers Kadence, ten, and Kyle, seven. They ride the same brushy, rocky country as the grown-ups, trot out before daylight and spend a full day in the saddle. Their folks are confident that they'll be ready to run that crew someday.

FALL 2017

Heart Mountain Homesteaders

A good life in postwar Wyoming. By Sheryl Lain

I grew up under Heart Mountain, its silhouette an emblem of Powell, a small farming town in northern Wyoming. They say a volcano erupted 50 million years ago unloosing a huge block of limestone that came to rest on top of older rock on the floor of the Big Horn Basin. This 8,000-foot boulder became a landmark for pioneer farmers who, eons later, called it Heart Mountain, named by the Crow Indians because they thought the mountain looked like a buffalo heart. When Heart Mountain broke loose, so did a million little rocks that rained down on the valley floor. Today, the big rock remains, but the little ones were picked up by homesteaders who moved them out of the way of plow blades.

Starting in 1947, soldiers returning from the Pacific and European theaters in World War II stood in line at the Park County Courthouse and considered themselves lucky to draw 160 irrigable acres in the new Heart Mountain Reclamation Project. My dad, along with the other winners, promised to reclaim this land. In the process, all of us—husbands, wives and children—claimed our pull-yourself-up-by-the-bootstraps legacy.

The homesteaders went to work to prove up on their places. Besides reclaiming the land, they repurposed the tarpaper barracks that housed war internees from the nearby relocation camp. After Japan bombed Pearl Harbor on Dec. 7, 1941, President Franklin Roosevelt signed an executive order to remove people who might support Japan in the war. Of the 120,000 Japanese moved inland, 10,000 came to Heart Mountain, creating in a finger snap

Dad and the other modern-day pioneers domesticated this wild place from its prior denizens, the rocks, rattlers, and sagebrush.

The author in front of the homestead house in 1954. AT TOP: Winter on Heart Mountain with barracks moved from the relocation camp in the foreground.

the third largest city in Wyoming.

The land for the camp was 46,000 acres between the towns of Cody and Powell, bordering the Shoshone River. Originally this area was part of the Heart Mountain Irrigation Project, which was part of the larger Shoshone Project, both overseen by the Bureau of Reclamation. The work on the project was halted due to the war.

The farmers drove their dusty pickups to the empty camp and either tore down their barracks or prepared to move it. My dad elected to take down his barracks board by board. Carefully the claws of his hammer gripped the nails and pulled them out of the one-by-sixes, nails that screamed resistance. The kids straightened those nails to reuse, for the times required frugality. These boards and nails were inside the walls of our home.

Dad subsidized his farm, unit 188, with a town job teaching agriculture classes to crop after crop of local farm boys. But most farmers didn't have the luxury of building a new house like Dad did. They pulled their barracks onto their land and lived inside, using their olive-green Army blankets to divide the 20-by-60-foot space into rooms. Today when I drive around the west end of Powell Valley, my experienced eyes recognize those barracks camouflaged with new siding, shingled roofs and additions.

After making homes, the work began in earnest. They ripped out sagebrush and leveled the fields for irrigation, using precious water from the Buffalo Bill Dam on the Shoshone River. Life-giving water would make this semiarid place green and productive. They put in fences and built barns, chick-

130

en coops, and corrals.

Dad worked in a frenzy, finishing project after project. He planted trees along the perimeter of the family yard, graveled the driveway, put in a black mailbox, and built a bridge across the irrigation ditch to the house. He bought a milk cow and a herd of registered Black Angus. He planted alfalfa and sugar beets. He plowed and prepared the ground for a kitchen garden.

Dad and the other modern-day pioneers domesticated this wild place from its prior denizens, the rocks, rattlers, and sagebrush. With sweat and ceaseless toil, these farmers followed the biblical injunction to have dominion over the earth, defined by new fences marking their farms.

Farm accidents were common. My mom got caught in a post-hole digger—the wind catching her long coat like a sail. Eagerly the alligator teeth of the post-hole digger snapped hold of the wool, tangling Mom in its jaws, grinding her into the ground. Dad shut down the tractor just in time to save her, though she took months to heal.

In spite of the work and the danger, the farm women considered their lives easier than the lives of their mothers. After all, only a generation before, their mothers used wringer washers while they had automatics. Their moms made lye soap from fat and ash while

The Bishopps, left to right, back row: Robert, Vivian and Sheryl. Front row: Corlienne, Glenya and Janell. LEFT: The tamed land yielded hay to feed cows and other critters. The John Deere tractor is pulling a hay wagon. BELOW: Moving barracks from the Heart Mountain camp.

they bought Tide or All in boxes from the grocery store. Their moms milked the cows, a division of labor that for some reason fell to men on our modern farms.

We were an unusual bunch, we Heart Mountain homesteaders. We believed that the work was worthy, even redeeming, something to meet eye to eye. We were unusual too because of our brand of fun. You've never seen fun when work is the staple of life. Every month—summer, winter, spring or fall—all us homesteaders gathered at the community club, another barracks scantily modernized to include a gas range and an inside toilet. We'd gather for potlucks. We ate chicken and roast beef and mashed potatoes whipped by hand to smoothness and crowned with someone's

golden homemade butter. We ate corn cut from the cob and preserved in Kerr jars—a reminder of autumn. We drank milk, non-pasteurized, and coffee from the tall, silver-colored urn. We ate homemade bread, warm and yeasty, and made-from-scratch cakes or rhubarb pies with lattice-top crusts. The recipe for green Jell-O salad was introduced at those feasts.

After dinner, Swede Olson, his homestead a few miles down the road, always played honky-tonk piano by ear, the four spindly legs of the stool bulging under his considerable weight jiggling in time to the music. We circled up for square dancing—the needle of the record player a little silver-pointed sliver we dared not lose in the cracks of the wood-plank floor. The kids chased around and whispered and giggled while the men exchanged ideas about what to raise next season, what price they'd gotten for the sheep. The women talked of garden seeds and Simplicity patterns.

Life on the last homesteads made an indelible imprint on all of us. We were different, from the boards of our houses and barns, to the work ethic woven into our DNA, to the nature of our fun.

Dad and other homesteaders reclaimed the semiarid land under Heart Mountain. They proved up, they used the boards of history to make their homes, and they gave us kids the legacy of hard work. When the sun set behind our talisman, Heart Mountain, we knew that after a good night's rest we'd be ready to tackle tomorrow's labor, ready to reap the rewards of the land that sustained us. ■

Sheryl Lain is a retired teacher from Cheyenne, Wyoming.

All in a Day's Play

Ranch kids learn to work young. Their play becomes practice for the skills they'll need when they get "older."

Challenge Issued at the Hoodoo

Near the south fork of the Shoshone River in northwest Wyoming, two-year-old Joshua Duncan wandered off, talking incoherently to the newly weaned Hoodoo Ranch calves. Curious, they gathered around and listened carefully.
© *Jeff Duncan, foreman, Hoodoo Ranch*

Buttercup Needs Help

Cassie Cash, age four, likes to do all the things adults do. Her grandmother, Debbi Rossi, was a horseshoer. The hoof pick she is using on her pony, Buttercup, was given to her by my boss, Scott Sumner, who had made it especially for her. She cherished it. She's getting ready to go to a jackpot roping. © Shannon Cash

"It's important kids grow up on a ranch. They learn to work. There is so much to life out here that city kids don't have the privilege of experiencing. If you didn't love animals and the solitude of country life, you wouldn't want to ranch. However, I do. And I've loved every minute of it."

—*Bunny Connell, 91, Wyoming*

My Turn!

My happy little cowgirl, Rilee, braves the rain to feed her bummer calves at the barn. © Kait Medart

Dean Thompson, 88

The poet of Lightning Valley.

There's no better memory for Dean Thompson than a beef or a hog hanging on the windmill tower in January, covered with a clean bedsheet. "Yes, I know," he says, "you're not supposed to freeze and thaw, but a round steak with ice crystals in the marbling is pretty good anyway." This memory, he says, sustained him through World War II rations and brought him home ready for the tough conditions of midcentury ranching in the Sandhills of Nebraska.

Dean was born in Sutherland in 1926, the grandson of homesteaders from Leon, Iowa. They lived near Flats, where his dad, Orville Dean Thompson Sr., carried the mail from Philbrook Valley to Sutherland, "three round-trips a week and 40 wire gates to open. Dad was always glad when there was a passenger."

He cannot remember a time when his entire family wasn't involved in the haying business. His mom drove the stacker team or ran a hay rake and later on became full-time cook for the crews in two sod-shanty haying camps.

When Dean and his siblings were young, "Mom had to stay with us, so Dad ran half a crew, mowing in the forenoon and stacking in the afternoon." In 1936, during the depths of the Depression, Orville bought the second John Deere tractor in the country, with double mowers, and started haying twice as much. "I was a proud kid of 11 when one of the hired men quit the second week of haying and my dad told me to 'get ready to go to work.'"

In the early days, their folks milked as many as 20 cows by hand. "They carried the milk from the barn to the house," Dean says, "the only place warm enough in winter to keep the cream separator. They extracted the cream by centrifuge, carried the skimmed milk back to

the barn and fed it to the bucket calves." Without a family car for trips to town, they'd send the cream with the mail carrier and ride horseback or take the team.

"I remember a little car with running gears that Dad rigged up that pulled easy. We traveled pretty light and fast with it. I fell off that rig one day and Mom was driving the team and ran right over me, got scared and backed across me again without

PHOTOS COURTESY THOMPSON FAMILY

CLOCKWISE FROM LEFT: Dean with Doris in 2014, "the love of my life. She and I look older now after nearly 67 years battling the elements and each other." ➤ *The Thompson kids share a horse, ca. 1957. From left: Brian, Barbara and Sandra.* ➤ *Dean at the sod house he helped build. "We dug up the buffalo skull when we were haying in a dry lake bed."* ➤ *The wedding photo in 1947.* ➤ *"This is the big soddie on Granddad's homestead, where my folks brought me when I was 19 months old. My paternal grandmother is with two of my uncles and one aunt."*

> **"My dad carried the mail from Philbrook Valley to Sutherland, three round-trips a week and 40 wire gates to open. Dad was always glad when there was a passenger."**

thinking, and it didn't hurt me at all."

The Sandhills have always been better suited to cattle ranching. "I can barely remember the days when we still made an effort at farming," Dean says, "although some of the neighbors struggled longer with that myth than Dad did. Still, we thought it was a big deal to be able to pick up a few buckets of corncobs out of the

hog pens to start a fire, rather than gathering cow chips."

Dean made it through school and found himself in the Pacific during World War II. "I was in the Philippines when they dropped the bomb on Hiroshima," he recalls, but soon after relocated to wartime offices in Osaka, Japan. At war's end, Dean came home to Nebraska, where he went back to haying, married Doris Tiffany, and in 1947 bought what is now the Stepladder Angus Ranch, where he and son, Bryan, raise and sell Angus bulls.

To keep himself amused, Dean writes ranch poetry, some of which has been set to music by his longtime friend, Lynn Myers. He also writes a column for the Arthur newspaper called "Between the Lines." ("You'd think they'd at least give me a subscription," he says wryly.) He and Doris have spent a significant amount of time in the Baptist Outreach Program for their church, which has taken them as far afield as Haiti and Ecuador.

How did they manage all this globe-trotting? Dean says with a smile, "Bryan stayed home and took care of the bulls."

—*Carolyn Dufurrena*

Glenn Plato, 89

How did I get in this mess?

"I'm a country boy, 20 years old," Glenn Plato remembers thinking. "How in hell did I get in this mess?"

It was May 7, 1942, and he was floating in the South Pacific, watching his ship, the USS Lexington, flounder after being crippled by two Japanese torpedoes. When given orders to abandon ship, Glenn and his shipmates dropped to the water. "I wasn't in any hurry because the place was loaded with sharks and I wanted them to get their fill," the 89-year-old remembers.

The sharks had been scared away by the blasts, so he and the others swam and floated, waiting to be rescued by other ships in the U.S. fleet. "The water was nice and warm," Glenn recalls. "I was a strong swimmer and it didn't bother me at all. We were in the water quite a while." They were retrieved by a Navy cruiser after about two hours.

The Lexington was sunk during the Battle of the Coral Sea, fought between the Imperial Japanese Navy and U.S. and Australian naval and air forces. Of the Lexington's 2,951-man crew, 216 died and 36 aircraft were lost. After the survivors were rescued, a U.S. destroyer fired torpedoes into the Lexington, preferring to sink it rather than allow its capture.

Glenn wanted to be a pilot or just be around airplanes, so he enlisted as a Navy aviation machinist mate in 1941 after working the nightshift in a Lakeview, Ore., mill. He had just finished boot camp and trade school when the Japanese bombed Pearl Harbor, triggering U.S. entry into World War II. He was sent to Hawaii, where he joined the Lexington.

Not long after that ship's sinking, his fortunes soared, literally, when he was accepted into Navy flight school. He flew a thousand hours for the Navy. "That's when my career

It was May 7, 1942, and he was floating in the South Pacific, watching his ship, the USS Lexington, flounder after being crippled by two Japanese torpedoes. When given orders to abandon ship, Glenn and his shipmates dropped to the water. "I wasn't in any hurry because the place was loaded with sharks and I wanted them to get their fill."

PHOTOS COURTESY GLENN PLATO

CLOCKWISE FROM TOP: Glenn Plato, front right, with mates, while assigned to the Army in New Caledonia during World War II. ➤ *Glenn during his early Navy years. He ended his military career with plenty of stripes.* ➤ *Glenn shows a photo of the USS Lexington being torpedoed and sunk to prevent it from being captured by the Japanese.*

really started," says Glenn, who shaped his WWII flying lessons into a career as a crop duster for southeastern Oregon farmers and ranchers.

When the war ended, he wasted no time returning to Lakeview and marrying Delpha, the 1941 Lake County Roundup Queen. "That happened the day I got out," he crows of his marriage to Delpha. "Once I got out of the service you might say that's

when my life began."

Glenn worked as the Lakeview airport manager, where he also taught flying lessons and began his 30-year crop-dusting career, eventually flying another 10,000 hours. "I did a lot of dusting in the desert," he says. "It was something I was good at, flying low."

He says that flying is a solitary thing that you learn how to do. "You get to see the world from a different perspective entirely. And it was interesting, flying and crop dusting in the country."

Glenn has done masonry and house construction and worked on the family ranch south of Lakeview. Delpha died in August 2010.

"When I look back, I've had a good life, varied experiences," he says with evident satisfaction. Then he adds with a grin, "Time flies."—*Lee Juillerat*

The Last Daughter of Hole in the Rock

Sara Ann Holyoak is the last living link to the generation of pioneers who settled Utah's red rock outposts. By Majorie Haun

Living as ordinary a life as any 91-year-old woman, it may be easy for Sara Ann Holyoak to minimize the extraordinary ties that link her to the establishment of Utah's remote red rock outposts nearly 150 years ago. As a baby, Sara Ann's father survived the bitter hardships that tested the Hole in the Rock pioneers, and her grandparents are memorialized today at the reconstructed settlement of Bluff Fort. Although her father died when she was an infant, the profuse and detailed family histories she and other relatives have amassed weave a vivid tapestry of remembrance. As the last daughter of the first Hole in the Rock pioneers, Sara Ann Holyoak has a remarkable story to tell.

The Hole in the Rock Legacy

Sara Ann's father, Albert "Daniel" Holyoak, was an infant when his parents, Henry and Sarah Ann, answered a mission call from their church to "settle the San Juan." In October of 1879 they willingly pulled up stakes, left Paragonah in western Utah, and endured a six-month journey that was estimated would take only six weeks. With two wagons, five young children, and a small herd of cattle, these strapping servants of the Lord headed east into Utah's terra incognita.

Sara Ann's grandmother drove a team of horses pulling an oversized wagon large enough to carry a stove. Baby Daniel was kept warm by the stove, as were the producers of their sustaining honey, a hive of bees. Grandfather Henry drove his own wagon pulled by a team of oxen, and their oldest boy, nine-year-old "Uncle" John, was tasked with driving the cows.

The most harrowing leg of the trek was through a narrow rock corridor known as the Hole in the Rock. The advance party scoured

At 91, Sara Ann enjoys the company of family who revel in her colorful tales of Moab's bygone years.

the area for an easier route, but the glorified chink in the canyon rim above the Colorado River was the only possible ingress. The men dug and blasted for several weeks to make the steep trail just wide enough for a wagon. The wheel hubs on the large wagon driven by Sara

> **The draft animals were so spent that the pioneers ended their journey short of their actual destination of Montezuma. To venture further would have killed the animals upon which their lives depended.**

Ann's grandmother carved deep horizontal grooves into the cliff faces as it descended down "the hole." After navigating the river and starting up the hill opposite the Hole in the Rock, the stay chain broke and threw Sarah Ann's wagon on its side, spilling the

bees and baby Daniel. Amazingly, Daniel was unhurt, but the bees had to be sacked up and the wagon taken apart and packed up the steep hill in pieces.

As the Hole in the Rock tested the courage of the men and women, the final hill leading out of the bottom-lands tested the mettle of the horses and oxen to their extremity. The draft animals were so spent, it is said, that the pioneers ended their journey short of their actual destination of Montezuma and settled in the area now known as Bluff. To venture further would have killed the animals upon which their lives depended.

The Upriver Settlement

The first settlement fell short of anything that could be called civilization. Bivouacked in the desert with scanty resources, some of the settlers, including the Holyoaks, opted to move 25 miles upstream on the San Juan River to Montezuma. There they found little relief from adversity. Repeated struggles with privation and the elements precluded building a house, so for four years they lived out of their wagons with little protection from the harsh climate. Year after year crops were flooded out by the river's spring surges. The final blow came in 1884 when the waterwheel they had labored to build was washed away and later found downriver. A measure of relief would come shortly thereafter when they were formally released from their settlement mission. Sara Ann's grandparents, with their remaining cows in tow and young children miraculously intact, headed north to Moab where several farms and ranches had already been established.

Cattleman, Husband, Father (in that order)

Daniel, Sara Ann's father, was four years old

FROM ABOVE: The Holyoak family at their cabin, ca. 1897. ➤Sarah and Henry Holyoak, ca. 1900. ➤Sara Ann's parents, Daniel Holyoak and Etholen Silver Dunn, ca. 1924. ➤Sara Ann, age one.

when the family arrived in Moab. There, he grew up to be a rancher, and during the early part of the century ran his cattle on the La Sal Mountains and along the Colorado River in a place called King's Bottom. A tenacious cattleman, Daniel and other Holyoaks built some of the most successful cattle operations in the area. Many of the local ranchers gave Daniel the responsibility of getting their cattle to market. That meant he had to drive them to either Thompson, 30 miles north of Moab, to be loaded on the train, or all the way to feedlots in Kansas. Sara Ann quips, "When he got those cows to market he was the richest man in town—until he had to pay all the other ranchers what he owed them."

Ranching consumed Daniel, and as he neared middle age he was also burdened with caring for his aging father. It wasn't until his mother died when Daniel was 46 that he considered marriage. He courted and took to wife one Etholen Silver Dunn, an "old maid" of 33 from Alamosa, Colo. "We found some of the letters they wrote to each other during their courtship," Sara Ann says. "We thought they were going to be love letters, but they were mostly just Daddy talking about his cows."

Marriage did not diminish his zeal for running cattle, and because Daniel was often out with his herds for two or three weeks at a stretch, Etholen was consigned to spates of loneliness and isolation. Nevertheless, Genevieve, their first child, was born in 1926. That same year, Daniel built a brick home for his family that many regarded as "before its time," since log cabins and line shacks were more typical in Moab. Sara Ann was born June 28, 1928, in that brick home.

As many ranchers did a hundred and more years ago, Daniel fairly worked himself to death. "The doctors told Daddy to sell his cattle because his heart couldn't take it," Sara Ann says. "But he wouldn't do that because that's how he provided for his family." The story goes that he had a premonition that he was going to die, and that the baby Etholen was carrying would be a redheaded son. Those events were realized. In 1929 at 49 years of age Daniel died, and baby Dan Holyoak was born just 22 days later. Etholen was left with three babies under the age of four to raise on her own.

The Moab Years

Ranching slowly gave way to simply surviving. Sara Ann's mother sold off the cattle and the land they no longer needed. Daniel, perhaps foreseeing his untimely death, had set aside a horse for each of the children. And Uncle John gave each child a calf to raise. In the following years, Sara Ann and her siblings helped to care for the little herd until one winter when they lost them all, save two milk cows, in a killing storm. With dreams of a ranching life dashed, Etholen and the children were left only to watch as others in the Holyoak clan carried on acquiring land and running cows for generations, into the present day.

Although the family owned a car, Etholen didn't drive, so walking was their first mode of transportation between home, school and work, and cantankerous horses their second. Swaybacked, easy to spook and hard to catch, Madame Queen was the family horse. Sara Ann speaks of the old mare's temperament: "I was riding on her bareback after Halloween one time, and there was a

Clockwise from above: Top row from left in "our swimming hole": Vera Holyoak, Sara Ann, Genevieve, Wynona, Louie Annalee and Shirley Anderson. Front row: Dan, Dale Holyoak and Billy Anderson. ➤ Daniel with Fred, ca. 1920. ➤ Sara Ann and little Dan, 1931. ➤ Central to the restored Bluff Fort is a replica of the waterwheel built by Sara Ann's grandparents, which now stands as a memorial to their hard-fought quest to settle the San Juan.

mask hanging on a neighbor's gate. That scared her, and she threw me off sideways. I couldn't find a way to get back on with no saddle or stirrups, so first I had to catch her, which took a while, then I had to lead her all the way back home."

Dan, the youngest and only son, had a Shetland pony named Cupie and together they made an ornery pair. Sara Ann says: "Me and Momma had to walk several miles to the grocery store, and on the way back, Dan came riding up beside us on Cupie. I said, 'Dan, why don't you let me ride for a little while, I'm awfully tired.'" She handed the groceries to Dan and climbed on.

"The minute I got on that pony, Dan slapped him real hard on the rear and off he went. I had never been on a running horse before and the louder I screamed, the harder he ran!"

Starstruck Girl

For the first half of the 20th century, Moab's key industries were ranching, mining, and moviemaking, a curious mix of the Old West and the Atomic Age. Sara Ann recalls: "One day I was working at the drugstore and sud-

denly I saw all these cows running up Main Street. I asked somebody what in the world was going on, and they told me they were just driving the cattle to Thompson." With Moab settled in a long narrow valley, the cowboys had no option but to drive their herds of thousands straight through downtown, hooves clattering on the wooden Colorado River bridge as they headed north.

The larger-than-life presence of John Wayne was a common reality in Moab through the 1930s and '40s. The region's raw beauty drew the attention of Hollywood directors long before it drew the millions of tourists now clotting its highways. The great John Ford was just one of many legendary filmmakers who used the area's variegated rock towers and quiet river canyons as settings for some of the greatest Westerns ever made. Back then, locals were often invited to the movie sets to act as extras or just to spend a day watching the actors and film crews at work. Always looming head and shoulders

above all was John Wayne.

During the World War II years, many doctors and men of all professions had been called to duty, so Sara Ann worked alongside her older sister, Genevieve, helping the nursing staff at the Moab Hospital. "I was working one evening and I heard somebody come into the outer office. I went to see who it was, and it was John Wayne."

Sara Ann continues to relate one of the most memorable events of her youth: "One of their trick riders had been hurt and was in the hospital, so John Wayne asked to see him. He didn't know where to find the boy, so I walked down the hall with him towards the room, and as I was walking I felt his arm around my shoulder!" Blushing like the comely girl she was 75 years ago, she says: "I thought to myself, 'Is this really happening, and how did I get to be so lucky?' After he left, I went into that trick rider's room and told him, 'Nobody, nobody ever put his arm around me like he did.'" Sara Ann confesses, "I didn't date that trick rider, but I did date some of those young movie actors and stunt riders back then."

An Honored Life

In 2018, Sara Ann turned 90, and in a ceremony at the reconstructed pioneer settlement of Bluff Fort she was honored as the last of her generation. Surrounded by younger generations of Holyoaks, she witnessed the dedication of a replica of the waterwheel her grandparents built before it washed downstream in 1884. Also in 2018, in a sad instance illustrative of the vexatious changes Moab has suffered in recent years, the brick ranch house her father built in 1926 was torn down, and a chain hotel was put up in its place.

Now in her 10th decade, Sara Ann looks at life through joyful eyes. As a mother and grandmother with a few "greats" following the title, family is everything to her, and the stories of past generations come alive in her plucky anecdotes. In her veins pumps the blood of true pioneers, and her life is a final, tenuous link connecting southern Utah's immutable past to an uncharted future. ∎

As a resident of San Juan County, Marjorie Haun is well acquainted with the many Holyoaks living throughout southern Utah. She is a freelance journalist specializing in rural issues, environmental activism and government run-amok.

Good Buddies

© Larry Turner

Ten-year-old Mia Davies adoringly places her arm around her grandfather, Stacy Davies, at Oregon's Roaring Springs Ranch. She is Stacy's eldest grandchild and the daughter of his eldest child, Zed, who is a U.S. Air Force (and 2012 U.S. Air Force Academy graduate) pilot in Europe. Mia was on a summer visit with her grandparents. Stacy manages the one-and-a-half-million-acre ranch outside Frenchglen.

Deep Roots

Snapshots from God's own cow country in Nebraska's Sandhills. By Beth Gibbons

Sod house of John R. Wysong and Emma Lucretia Robertson Wysong, shown here with their daughter, Revia Jane. Revia Jane later married Rex Leroy Chase and continued the ranch operation. It is now run by their son, Kenneth, and granddaughters Julie and Patricia.

The first cattle arrived in Nebraska in response to the 1878 settlement of some 5,000 Sioux Indians on the Rosebud Reservation in South Dakota. The federal government had promised to provide beef for the Indians on the reservation, so large herds were established on the good grazing land along the Niobrara River. Cherry County was founded in 1883 and named after Fifth Cavalry Lieutenant Samuel A. Cherry, who had been murdered near the Army's Fort Niobrara two years before. It didn't take long for word to spread that "the heart of cattle country" was open to anyone with a hankering to homestead and try a hand at ranching. Among those with enough guts was John Remley Wysong. A Missourian who arrived on horseback in 1895, Wysong quickly patented his government-granted 160 acres.

In the Sandhills, a quarter-section of land doesn't make for much of a cattle ranch, so it's not too surprising that Wysong took obvious interest in the arrival, four years later, of Miss Emma Lucretia Robertson from Ohio. Emma had acquired

It didn't take long for word to spread that the county that now liked to call itself "the heart of cattle country" was open to anyone with a hankering to homestead there and try a hand at ranching. Among those with enough guts was John Remley Wysong, a Missourian who arrived in southeastern Cherry County in 1895 on horseback and quickly patented his government-granted 160 acres with a homestead.

a section adjoining his. That might not have been the reason John and Emma married five years later, but soon enough they had two children, John Zachariah and Revia Jane, and their two quarter-sections

had been joined. This was the start of the CC Ranch, today encompassing some 12,000 acres.

The CC Ranch boss today is Kenneth Chase, whose father, Rex Leroy Chase, came to the area in the early 1900s. He married Revia Jane in 1927. Rex and Revia Jane expanded their holdings to some 3,200 acres, Ken recalls, and Revia bore six children along the way—Ken's brothers John, Garold and Robert, and sisters Emma and Ernia. Ken himself met and married Carolyn Babbitt from Ainsworth, Neb., in 1957, and they raised their three daughters—Julie, Jacqueline, and Patricia—as Ken expanded the ranch to its present size. His wife, Carolyn, died from Lou Gehrig's disease in 1997, so today Julie, Patricia and her husband, Terry Shoemaker, help their father run the ranch. Patricia and Terry live in the house that Rex and Revia Jane built in 1947.

Ken remembers the devastating and uncontrollable wildfire that swept the Sandhills in March 1943, which burned some of the ranch's good grassland. Ken

LEFT: John Remley Wysong and Emma Lucretia Robertson wedding day, Dec. 4, 1904.
CENTER: Rex Leroy Chase and Revia Jane Wysong wedding day, March 18, 1927.
RIGHT: Wedding picture of Kenneth Harold Chase and Carolyn Elaine Babbitt, June 7, 1957.

Patricia Chase Shoemaker, Julie Chase, their father, Ken Chase, and Jackie Staudenmaier at their home near Elsmere, Nebraska.

still remembers that his ementary school was dismissed during the fire so that the older kids could help their families fight it. Fortunately, today the effects of that fire have disappeared.

Before the Second World War, Ken says, "We did all our work with horses." Today they keep just four saddle horses. Like most modern ranches, the CC stables are more full of machinery these days. "Our first tractor was a John Deere D," he says. A succession of Farmall tractors followed after the war and they continue to use some horses on the place. "The horses are for when the snow is too deep for the tractors."

As anyone who lives in Cherry County will tell you, the snow can get plenty deep. The CC Ranch still flourishes, and John and Emma Wysong's descendants keep a busy schedule fencing, checking 30 windmills, and putting up hay to feed the livestock, as they grow their roots ever deeper into "God's Own Cow Country." ■

Beth Gibbons is a writer, teacher and grandmother who lives in Crawford, Nebraska.

ABOVE: CC Ranch, 2007. Notice the addition of many trees for beauty and protection. Roads around the buildings increased ease of access. BELOW: Rex and Revia Jane's home, built in 1948. Note windmill, a source of water for domestic use and animals, and the wind charger, the source for electricity for the ranch. Trees did not grow until there was a more generous water supply.

NEVADA . Spring 2021
Marge Prunty, 94

I love ranching and this way of life.

For her 94th birthday, the matriarch of two of the West's oldest ranching families enjoyed a well-marbled prime rib eye for dinner. With the exception of maybe a year or so of mother's milk, Marge Bieroth Prunty has been a lifelong consumer of red meat. "Beef was practically part of every meal almost every day. We raised our own, but sometimes we would break it up a bit with store-bought beef."

Marge was born Aug. 25, 1926, in Elko, Nev., already a third-generation cattle rancher. Her grandfather, Sam Bieroth, an emigrant from Germany, had started the ranching tradition in the 1880s on McDonald Creek in northern Elko County. Her father, Hugh, ranched on California Creek, a few miles from Mountain City.

Clockwise from left: Grandfather Sam Bieroth and family at the McDonald Creek Ranch in the early 1920s. Marge's father, Hugh, is behind the lady in long white dress. Her mother is to Hugh's right holding Marge's brother George. Marge is still a twinkle in her daddy's eye at this point. ➤ Marge Prunty, 94 years old in August 2020, on her four-wheeler with her dog, Kip. ➤ Marge and granddaughters Kyla, left, and Becky (Dick's daughters and heirs to the ranch). ➤ Shorty Prunty mid-1990s. ➤ Marge and Shorty's wedding photo in 1948. ➤ Marge with sons Gary, left, and Dick.

ranch, teaching school, and cooking for hunters and hay crews, Marge also raised two sons, Gary and Dick!

"I am delighted this home ranch has been in the family for over 100 years. It's owned and run now by my nephew, Dennis Beiroth. My dad was the first in Nevada to introduce Black Angus cattle. Prior to that most cowboys in Nevada ran Herefords."

Mountain City became a boomtown when one of the richest copper deposits ever found happened there in the 1930s—the famous Rio Tinto mine. Demand for produce skyrocketed and her dad started a dairy.

"We kids all worked hard, before sunup milking the dairy herd and then my brother drove us to school and on the way we delivered the milk. And after school we did it again."

After attending school in Rio Tinto and graduating from Elko High School, Marge got her degree in education from the University of Nevada in 1947. "My first teaching assignment was at the Indian reservation at Owyhee, and I still have close friends there. In fact, Indians were our favorite workers at the ranch."

It was not always work. "We loved to go to dances and they had big ones at North Fork

nearly every week." She met a tall stocky cowboy at one of those dances who went by the misnomer "Shorty." Frank Prunty was another third-generation rancher and he and Marge were married on July 15, 1948, thus merging two great ranch families.

Marge and Shorty soon bought out her in-laws and took over the Charleston ranch, which has been Marge's beloved home ranch for 72 years. "Our first winter there was the year of 'The Great Haylift' when the Air Force dropped hay to stranded cattle. The snow and cold were terrible."

Entrepreneurs both, besides running the ranch Shorty supplied stock for rodeos and became a major stock contractor from 1948 to 1968 while Marge educated children.

"I taught for 17 years, in Owyhee, North Fork, Charleston, Elko, and a couple of years in Jerome, Idaho. I am one of the last of the one-room schoolteachers with multiple grades in one room."

Shorty was a licensed master hunting guide and Marge was the cook. "I got up at four a.m., cooking for dozens of hunters and guides for decades." In addition to running a

jobs paid the bills, Marge's real pride and joy are the Prunty horses. Besides Hall of Fame quality rodeo stock, Prunty horses with their proven champion bloodlines are internationally known. "Our horses are in demand around the world. We just sold one and shipped it to Italy for Heaven's sake!"

Shorty passed away in 1997 and although devastated by the loss, Marge carried on. Today the ranch remains a family operation, with son Gary and granddaughters Becky (Prunty) Lisle and Kyla (Prunty) Rianda slowly taking over its management.

Marge's contributions to the livestock industry, horse breeding and the western way of life has not gone unrecognized. She was featured in the Winter 2004 issue of *RANGE* magazine [see "Horse Women of the Diamond A" at rangedex.com] and was grand marshall for the 2005 Silver State Stampede, Elko County's fair and rodeo.

And with a pure dash of western horse sense, Marge laughs and asks, "If red meat is supposedly so bad for you, why the hell am I still here?"—*Ira Hansen*

A Very Sharp Dresser

© Cynthia Baldauf

Kaleb Wayne Donker, with bent ears and one spur, always dresses to kill for homeschool.

On the Ruby Ranch, 80 miles from town in the Big Hole Valley, Montana, four-year-old Kaleb Wayne Donker gets dressed for the day. Every day. Vest, suit jacket, string tie, badge. "Call me 'Tricky,'" he says.

Kaleb and his brothers—J.D., age nine; Garrett, age seven; and baby Vade—are growing up the old way. Parents, Nick and Jessi Donker, live with their four boys in a tiny cabin on a vast outfit. Nick rides for the ranch, trains horses, and feeds with a draft-horse team in the winter. Jessi runs the homeschool.

The three older boys are all replicas of their father. Kaleb trains his aunt's 30-year-old miniature stallion, which rules the draft horse pen with flattened ears. "Once he lets me catch him, he follows me everywhere," says Tricky. "I want to be a cowboy and a horse trainer like my dad!"

This Is How to Stuff
Your Cheeks, Grandpa!

Copperopolis cowboy Steve Wooster smiles as his almost two-year-old grandson, Lincoln Dell'Orto, gorges during the lunch break at a winter calf doctoring, marking and branding in the Sierra Foothills. For decades, three to four generations of Amador and Calaveras cowboys and ranchers, from newborns to great-grandparents, have shared chores at each others' ranches. © Larry Angier